Contents

Skills Guidance

Questions & Answers

■ About this book

This guide is written for both OCR specifications (A and B) for AS and A-level chemistry. Its aim is to help you understand the different practical skills that you are expected to develop throughout the course. These skills are assessed throughout the written examinations and also through the practical endorsement.

Chemistry is a practical subject and practical skills are fundamental to understanding the nature of chemistry. Chemistry gives learners many opportunities to develop the fundamental skills needed to collect and analyse empirical data. In the written papers the following written practical skills are assessed:

- **Planning** — including selection of appropriate apparatus, equipment and techniques for the proposed experiment
- **Implementing** — including detailing how to use a wide range of practical apparatus and techniques properly, and presenting data and observations in an appropriate format
- **Analysis** — including reaching valid conclusions by processing, analysing and interpreting experimental results
- **Evaluation** — including evaluating results, identifying anomalies, limitations and calculating percentage errors and uncertainties.

Questions are included throughout this book to help you develop these skills.

Practical work leads to many different types of calculation, many of which require the use of **significant figures**. The section on pp. 7–9 briefly recaps on how to use significant figures correctly. You may also find it useful to refer back to this section when you are trying questions later in the book. Other mathematical skills are identified in the practical activity groups.

The **Practical Activity Groups** (PAGs) section reflects the OCR requirement that you should acquire competence and confidence in a variety of practical, mathematical and problem-solving skills, and in handling apparatus competently and safely. You should be able to design and carry out both the core practical activities and your own investigations. Practical work is central to any study of chemistry. For this reason, the specification includes 12 PAGs that form a thread linking theoretical knowledge and understanding to practical scenarios. In this section eleven of the PAGs are discussed in detail, with worked examples included to allow you to think more deeply about why different steps are carried out. Working conscientiously through these examples will give you practice at answering the sort of question you will be asked in your examinations. For the AS examination you need only be familiar with PAGs 1–5. The twelfth PAG is not covered here, as it is a research-based activity.

The **Questions & Answers** section pulls together the other sections through a range of questions based on the type of question that will be asked in examination papers to test practical knowledge. Answers are provided as well as comments to help you to understand the way to approach each question.

AS/A-LEVEL YEARS 1 AND 2

STUDENT GUIDE

OCR

Chemistry A and B

Practical assessment

Nora Henry

Series editor: David Scott

HODDER
EDUCATION
AN HACHETTE UK COMPANY

Hodder Education, an Hachette UK company, Blenheim Court, George Street, Banbury, Oxfordshire OX16 5BH

Orders

Please contact Hachette UK Distribution, Hely Hutchinson Centre, Milton Road, Didcot, Oxfordshire, OX11 7HH

tel: 01235 827827

e-mail: education@hachette.co.uk

Lines are open 9.00 a.m.–5.00 p.m., Monday to Friday. You can also order through the Hodder Education website: www.hoddereducation.co.uk

© Nora Henry 2017

ISBN 978-1-4718-8564-8

First printed 2017

Impression number 6

Year 2021

This guide has been written specifically to support students preparing for the OCR AS and A-level Chemistry A and B examinations. The content has been neither approved nor endorsed by Edexcel and remains the sole responsibility of the author.

Cover photo: Ryan McVay/Photodisc/Getty Images/Professional Science 72

Typeset by Integra Software Services Pvt. Ltd, Pondicherry, India

Printed in Dubai

Hachette UK's policy is to use papers that are natural, renewable and recyclable products and made from wood grown in well-managed forests and other controlled sources. The logging and manufacturing processes are expected to conform to the environmental regulations of the country of origin.

The purpose of this book is to help you understand different practical skills in chemistry, and enable you to answer A-level papers, but don't forget that what you are doing is learning chemistry. The development of an understanding of chemistry can only evolve with experience, which means time spent thinking about chemistry, applying it to unfamiliar situations and solving problems. This book provides you with a platform for doing this.

If you are reading this, you are clearly determined to do well in your examinations. If you try all the knowledge checks, worked examples and the questions in the Questions & Answers section before looking at the answers, you will begin to develop understanding and the necessary techniques for answering practical examination questions. As the answers to the worked examples are an integral part of the learning process, they appear immediately after the questions. You are recommended to cover up the answers before attempting the question — this will improve your ability to answer similar questions in future. If you 'cheat' by looking up the answers first, you are only cheating yourself!

Thus prepared, you will be able to approach the examination with confidence.

■ Health and safety

Whenever experiments and investigations are carried out in the laboratory, general safety rules which you must follow are:
- Wear eye protection at all times.
- Take care when handling hot apparatus.
- Ensure there are no naked flames in the laboratory when using flammable substances.
- Dispose of chemicals as directed — take particular care in disposing of organic chemicals.

When planning an investigation you need to decide if the experiment is safe, by carrying out a **risk assessment**.

A good risk assessment includes:
- a list of all the hazards and why they are hazardous
- a list of the risks (things that you do in the experiment) which might result in danger
- suitable control measures you could take which will reduce or prevent the risk

You must recognise the Control of Substances Hazardous to Health (COSHH) hazard warning signs shown in Figure 1.

A **risk assessment** is a judgement of how likely it is that someone might come to harm if a planned action is carried out.

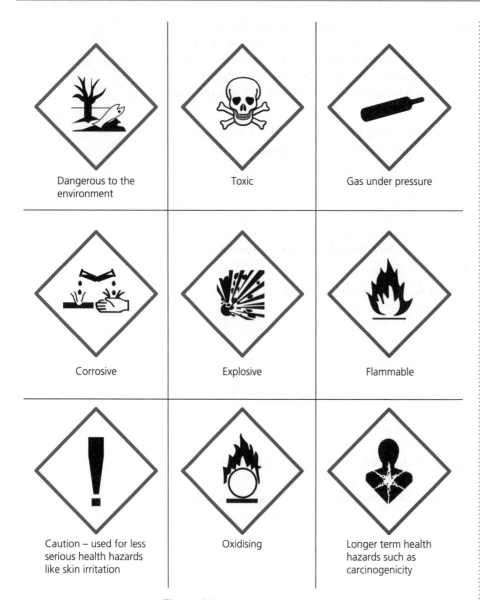

Figure 1 Hazard warning signs

In examinations you may encounter questions about safety applied to particular experiments, and examples of these are included throughout this book.

Hazard and risk

All practical tasks described in this guide should be risk assessed by a qualified teacher before being performed either as a demonstration or as a class practical. Safety goggles and a laboratory coat or apron must be worn where it is appropriate to do so. The author and the publisher cannot accept responsibility for safety.

Skills Guidance

■ Significant figures

Significant figures are those numbers that carry meaning and contribute to the number's precision. The first significant figure of a number is the first digit that is not a zero. In chemistry we quote values to a limited number of significant figures, as we cannot be sure of the exact value to a greater number of significant figures.

The rules for significant figures are:

1   Always count non-zero digits. For example, 21 has two significant figures and 8.923 has four.

2   Never count zeros at the start of a number (leading zeros) even when there is a decimal point in the number. For example, 021, 0021 and 0.0021 all have two significant figures.

3   Always count zeros which fall between two non-zero digits. For example, 20.8 has three significant figures and 0.00103004 has six significant figures.

4   When a number with no decimal point ends in several zeros, these zeros may or may not be significant. The number of significant figures should then be stated. For example: 20000 (to 3 s.f.) means that the number has been measured to the nearest 100 while 20000 (to 4 s.f.) means that the number has been measured to the nearest 10.

The rules are best illustrated with some examples:

■   34.23 has four significant figures — always count non-zero digits.

■   6000 has no decimal point and ends in several zeros, so it is difficult to say if the zeros are significant. With zeros at the end of a number, the number of significant figures should be stated.

■   2000.0 has a decimal place, hence it has five significant figures.

■   0.036 has two significant figures — never count the zeros at the start of a number even when there is a decimal point.

■   3.0212 has five significant figures.

In calculations you should round the answer to a certain number of significant figures.

The rules for rounding are:

■   If the next number is **5 or more**, round up.

■   If the next number is **4 or less**, do not round up.

Worked example

Calculate the value of 7.799 g – 6.250 g and give your answer to three significant figures.

Answer

Your calculation should yield 1.549 g. If the answer is given to three significant figures this would be 1.55 g, because the digit 9 is greater than 5.

When combining measurements with different degrees of accuracy and precision, *the accuracy of the final answer can be no greater than the least accurate measurement.* This means that when measurements are multiplied or divided, the answer can contain no more significant figures than the least accurate measurement.

Worked example

In a titration 20.5 cm^3 of 0.25 mol dm^{-3} sodium hydroxide solution reacts with 1.2 mol dm^{-3} hydrochloric acid. How many significant places should you give your answer to, when calculating the volume of hydrochloric acid required to neutralise the sodium hydroxide solution?

Answer

It is useful to look at all the data and write down the number of significant figures in each measurement. See Table 1.

Table 1

Measurement	Number of significant figures
20.5 cm^3	3
0.25 mol dm^{-3}	2
1.2 mol dm^{-3}	2

The least accurate measurement is to two significant figures, hence when you calculate your answer you must give it to two significant figures.

Significant figures and standard form

For numbers in scientific form, to find the number of significant figures ignore the exponent (*n* number) and apply the usual rules.

For example, 6.2090×10^{28} has five significant figures and 1.3×10^2 has two significant figures.

The same number of significant figures must be kept when converting between ordinary and standard form. For example:

- 0.0050 mol dm^{-3} = 5.0×10^{-3} mol dm^{-3} (2 s.f.)
- 40.06 g = 4.006×10^1 (4 s.f.)
- 90.0 g = 9.00×10^1 g (3 s.f.)
- 0.01070 kg = 1.070×10^{-2} (4 s.f.)

The number 260.99 rounded to 4 s.f. is 261.0

to 3 s.f. is 261

to 2 s.f. is 260

to 1 s.f. is 300

Using standard form makes it easier to identify significant figures. In the example above, 261 has been rounded to the two significant figure value of 260. However, if seen in isolation, it would be impossible to know whether the final zero in 260 is significant (and the value to three significant figures) or insignificant (and the value to two significant figures). Standard form, however, is unambiguous:

- 2.6×10^2 is to two significant figures
- 2.60×10^2 is to three significant figures

Worked example

Convert 0.002350 to standard form, ensuring that you retain the correct number of significant figures.

Answer

Step 1: note how many significant figures are present in 0.002350. Remember that the zeros after the point are not significant, so there are four significant figures.

Step 2: to write the number in standard form, write the four significant figures, with a decimal place after the first number, and then write '$\times 10^n$' after it.

2.350×10^n

Step 3: count how many places the decimal point has moved to the right and write this value as the n value. The n is negative because the decimal point has moved to the right instead of the left.

$0.002350 = 2.350 \times 10^{-3}$

> **Tip**
>
> Many calculations will ask you to give your answer for example to a certain number of significant figures *and* in standard form.

■ Practical activity groups

PAG1 Moles determination

It is possible to experimentally determine the amount in mol of a substance which reacts or is produced in a chemical reaction, and use this information to determine formula or relative formula masses of substances. Two ways of determining moles are by measurement of mass or measurement of volume of a gas.

Measurement of mass

The **empirical formula** for magnesium oxide can be determined by taking accurate measurements of the mass of a sample of magnesium, heating in air to form magnesium oxide, and then accurately measuring the mass of the product. The assembled apparatus is shown in Figure 2.

> **Empirical formula** gives the simplest ratio of atoms present in a formula.

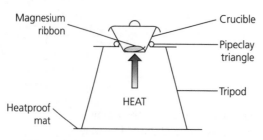

Figure 2 Formation of magnesium oxide

The method for the experiment is:

■ Measure the mass of a crucible and lid using a balance.
■ Add some magnesium filings, and measure the mass of the crucible, lid and magnesium.
■ Place the crucible on a pipeclay triangle and heat strongly, raising the lid at intervals to allow air into the crucible.
■ Allow the crucible to cool and weigh crucible, lid and contents.

Some sample results are given in Table 2.

Table 2

Mass of crucible + lid/g	25.85
Mass of magnesium + crucible + lid/g	26.45
Mass of magnesium oxide + crucible + lid/g	26.85

To calculate the formula of magnesium oxide, the mass measurements are used to calculate the amount in mol of each element present. See Table 3.

Table 3

Element	Magnesium	Oxygen
Mass/g	26.45 – 25.85 = 0.60	26.85 – 26.45 = 0.40
Divide by molar mass/g mol^{-1}	0.60/24.3	0.40/16.0
Amount/mol	0.0247	0.025
Divide by smallest amount to find ratio	0.99	1.00
Scale or round if necessary to find simplest ratio	1	1
Empirical formula	MgO	

Sources of error in this experiment include that not all of the magnesium is reacted, and when the lid is lifted, some of the product magnesium oxide may escape; this will result in a larger magnesium : oxygen ratio.

If **hydrated** compounds are heated they lose water of crystallisation (Figure 3) so their mass decreases and the anhydrous compound is formed. The decrease in mass is the mass of water lost. By heating the hydrated substance and taking mass measurement the degree of hydration can be determined.

Knowledge check 3

During heating the contents of the crucible were broken up with a spatula. What was the purpose of this?

A **hydrated** compound contains water of crystallisation, which is water that is chemically bonded in the crystal structure.

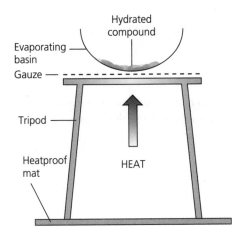

Figure 3 Heating to remove water of crystallisation

The assembled apparatus is shown in Figure 3. The method is:

■ Weigh crucible/evaporating basin.

■ Add hydrated solid and weigh crucible/evaporating basin with solid in it.

■ Heat to constant mass — this means heat for a few minutes, cool then weigh and repeat *until the mass does not change.*

Observations in this practical may include the crystalline solid changing to a powdery (anhydrous) solid and you may see steam given off or condensing on side of basin. There may be a colour change: for example hydrated copper(II) sulfate changes from blue to white.

Worked example

In an experiment to find the mass of water of crystallisation in hydrated magnesium sulfate, $MgSO_4.xH_2O$, the following results were obtained:

mass of empty evaporating basin = 12.73 g

mass of evaporating basin + hydrated magnesium sulfate = 13.96 g

mass after heating for 5 minutes = 13.56 g

mass after heating for 10 minutes = 13.33 g

mass after heating for 13 minutes = 13.33 g

Find the value of x in the hydrated magnesium sulfate.

Answer

The results show that the solid was heated to constant mass and all the water of crystallisation was removed so only anhydrous magnesium sulfate is left in the basin at the end. To calculate the mass of anhydrous magnesium sulfate, subtract the mass of the evaporating basin:

13.33 − 12.73 = 0.60 g

Then calculate the number of moles of anhydrous magnesium sulfate $MgSO_4$ ($M_r = 120.4$):

$$\text{moles} = \frac{\text{mass}}{M_r} = \frac{0.60}{120.4} = 0.005$$

To find the mass of water of crystallisation removed, you need to subtract the mass of the evaporating basin plus the anhydrous magnesium sulfate, from the mass of the basin plus the hydrated magnesium sulfate:

$$13.96 - 13.33 = 0.63\,\text{g}$$

Then calculate the number of moles of water ($M_r = 18.0$):

$$\text{moles} = \frac{\text{mass}}{M_r} = \frac{0.63}{18.0} = 0.035$$

When working out the value of x you are really working out the ratio of anhydrous salt to water. A convenient way of doing this is to use a table like Table 4.

Table 4

	$MgSO_4$	H_2O
Moles	0.005	0.035
Ratio (divide by smallest number of moles, 0.005)	$\frac{0.005}{0.005} = 1$	$\frac{0.035}{0.005} = 7$
Formula	$MgSO_4.7H_2O$	

Uncertainty in mass measurements

If a balance measures to one decimal place, the mass will normally have an uncertainty of $\pm 0.1\,\text{g}$. If a balance measures to two decimal places, the mass will normally have an uncertainty of $\pm 0.01\,\text{g}$.

Percentage error is calculated using:

$$\textbf{percentage error} = \frac{\textbf{uncertainty}}{\textbf{quantity measured}} \times \textbf{100}$$

Table 5 compares the percentage error in weighing 1 g using both balances.

Table 5

Balance to one decimal place	Balance to two decimal places
% error = $\frac{0.1}{1} \times 100 = 10\%$	% error = $\frac{0.01}{1} \times 100 = 1\%$

The *more* decimal places a balance reads to, the *smaller* the percentage error.

When weighing a *smaller* mass, the percentage error is *more significant*. Weighing 0.1 g on a balance to two decimal places gives a percentage error of $(0.01/0.1) \times 100 = 10\%$ compared to 1% when weighing the larger mass of 1 g.

Tip

To reduce the percentage error in the mass of water removed in this experiment, record the mass to three decimal places or use a larger mass of hydrated crystals.

Note that if two mass measurements are taken, as in many experiments, then the formula for the percentage uncertainty is then:

$$\text{percentage uncertainty} = \frac{2 \times \textbf{uncertainty in each measurement}}{\textbf{quantity measured}} \times 100$$

Knowledge check 6

In an experiment to find the mass of water lost on heating a solid, the following results were obtained:

mass of crucible + solid before heat = 25.45 g; uncertainty = 0.005 g

mass of crucible + solid after heat = 24.21 g; uncertainty = 0.005 g

Calculate the mass of water lost and the percentage uncertainty in this value.

Worked example

To determine the empirical formula of an oxide of titanium, some titanium metal was heated in a stream of oxygen as shown in Figure 4.

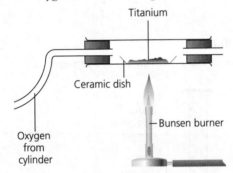

Figure 4 Heating titanium in a stream of oxygen

a Describe a test that could be carried out, before the cylinder was used, to prove that the gas in it was oxygen.

b What weighings would be made before heating to determine the mass of titanium used?

c In this reaction the titanium could be a solid lump or powdered. State and explain if there is any advantage in using titanium powder.

d The ceramic container and its contents are repeatedly weighed, heated, reweighed and heated. Suggest and explain the trend in expected results.

e What safety precautions should be taken in this experiment?

f Water vapour reacts with hot titanium to produce titanium oxide and hydrogen. Suggest how you could modify the apparatus to remove any traces of water vapour from the oxygen supply.

g In this experiment 4.8 g of titanium was oxidised to form 8.0 g of titanium oxide. Deduce the empirical formula of the titanium oxide.

→

Answers

a Oxygen relights a glowing splint.

b The mass of container and mass of container and titanium should be recorded.

c The powder would react faster.

d The mass would increase due to titanium oxide forming and eventually constant mass should be reached when all the titanium has reacted.

e Allow the container to cool before weighing and so prevent burns.

f Pass the oxygen through a u-tube, containing a drying agent (e.g. anhydrous sodium sulfate) before entering the test tube.

g moles of titanium = $\frac{4.8}{47.9}$ = 0.10

mass of oxygen = 8.0 − 4.8 = 3.2 g

moles of oxygen = $\frac{3.2}{16.0}$ = 0.20

ratio Ti : O = 0.10 : 0.20

formula is TiO_2

Measurement of volume of gas

When magnesium is reacted with excess dilute sulfuric acid solution to produce hydrogen, the apparatus in Figure 5 can be used to measure the volume of hydrogen produced.

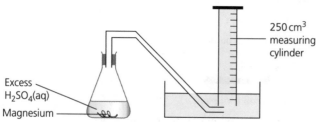

Figure 5 Collection and measurement of volume of gas

To start the reaction magnesium is dropped into the flask and the stopper quickly replaced.

There are some errors in this method:

1 Some gas may escape due to the time lag between adding the magnesium and replacing the bung, resulting in a decrease in the measured volume of gas.

To reduce this error, the magnesium could be suspended by a string above the acid and the stopper loosened just enough to release the thread, dropping the magnesium into the acid. Or alternatively, the magnesium could be placed in a small tube in the conical flask (Figure 6), acid added, bung replaced and then the flask swirled to mix the reactants.

Tip

Note that if this experiment is carried out using magnesium carbonate and acid, carbon dioxide gas is produced. A different error may occur in this case — some of the gas produced may dissolve in water. To reduce this error, use a gas syringe (Figure 6).

2   When the stopper is replaced, the volume of the bung displaces the same volume of air into the measuring cylinder, increasing the volume. Again the alternative methods above reduce this error.

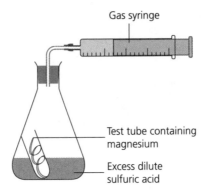

Gas syringe

Test tube containing magnesium

Excess dilute sulfuric acid

Figure 6 Using a gas syringe to collect a gas

In this experiment 0.12 g of magnesium was reacted with excess acid and 120 cm³ of hydrogen gas were produced. These values can be used to calculate the relative atomic mass of magnesium (A_r).

$$\text{amount in moles of H}_2 = \frac{\text{volume (H}_2)}{24\,000} = \frac{120}{24\,000} = 0.005\,\text{mol}$$

$$Mg + H_2SO_4 \rightarrow MgSO_4 + H_2$$

The ratio is 1 : 1 so moles Mg = 0.005 mol.

$$A_r(Mg) = \frac{\text{mass}}{\text{moles}} = \frac{0.12}{0.005} = 24.0$$

PAG2 Acid–base titration

Making up a standard solution

To prepare a **standard solution**, an accurately known mass must be dissolved in deionised water and then the volume made up to an accurate volume using a volumetric flask.

When preparing a standard solution a chemical should be used that:
- does not absorb or lose moisture to the environment
- has an accurately known relative formula mass, so that the number of moles dissolved can be determined — a hydrated salt is not used for a standard solution
- is very pure
- has a relatively high relative formula mass so that weighing errors are minimised

The method to prepare a standard solution is:
- Weigh out an accurate mass of a solid in a clean dry beaker.
- Add enough deionised water to dissolve the solid, stirring with a glass rod.
- Transfer the solution with rinsing to a 250 cm³ volumetric flask using a funnel.

Tip

Note that most gas syringes hold 100 cm³ of gas. So before carrying out an experiment, you need to check by moles calculation that the mass of solid used produces less than 100 cm³ of gas. The example with 0.12 g of magnesium produces 120 cm³ of gas, so the mass of magnesium must be halved before using a gas syringe, and a more accurate balance used.

Knowledge check 7

In this experiment two measurements were taken: mass of magnesium = 0.12 g and volume of hydrogen = 120 cm³. Calculate the percentage error in both readings if the balance uncertainty is ±0.01 g and the measuring cylinder uncertainty is ±1 cm³.

A **standard solution** is a solution for which the concentration is accurately known.

Knowledge check 8

Why is sodium hydroxide not a suitable solid for making up a standard solution?

- Rinse the beaker and glass rod with deionised water and pour the rinsings into the volumetric flask.
- Make up to the mark by adding deionised water until the bottom of the meniscus is on the mark.
- Stopper the flask and invert to mix thoroughly.

See also Figure 7.

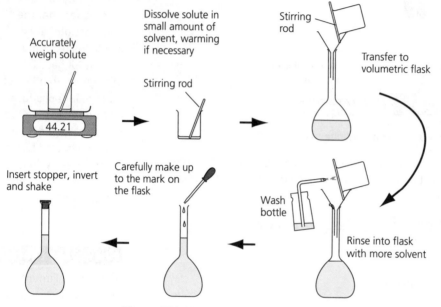

Figure 7 Making a standard solution

Titration method

The titration procedure is summarised as follows (see Figure 8):

- Rinse the burette with the solution you are going to fill it with. Discard the rinsings and fill the burette.
- Rinse a pipette with the solution you are going to pipette into the conical flask. Using a pipette and pipette filler transfer $25.0\,cm^3$ of this solution into a conical flask.
- Add 2–3 drops of a suitable indicator to the conical flask.
- Add the solution from the burette, with constant swirling until the indicator just changes colour. This is a trial titration.
- To reduce the effect of random error on titration results, repeat the titration to achieve 2–3 **concordant** results, adding the solution dropwise near the end point.
- Calculate the mean titre from concordant results.

> **Tip**
> When making up a standard solution, a larger mass of solid dissolved gives a smaller weighing error, so it will be a more accurate solution.

> **Tip**
> Measuring cylinders have limited accuracy and are not appropriate for use in titrations when volumes to one decimal place are needed.

Concordant titre results are those that agree to within $0.1\,cm^3$. Titration results should be recorded to two decimal places with the last figure either '0' or '5'.

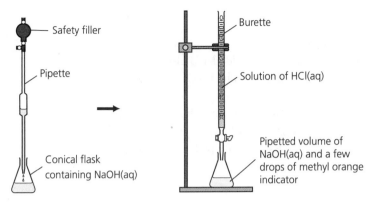

Figure 8 The apparatus used for a titration

Titration results should be recorded in a table. To calculate the mean titre, choose titres that are concordant (agree to within $0.1\,cm^3$). Where this is not possible, the two titres that have the closest agreement should be used. Readings that are not concordant are often called outliers, for example the value 22.85 in Table 6.

Table 6

	Trial	1	2	3
Final burette reading/cm³	22.90	45.50	22.85	45.55
Initial burette reading/cm³	0.00	22.90	0.00	22.85
Titre/cm³	22.90	22.60	22.85	22.70
Titres used to calculate mean		✔		✔
Mean titre		22.65cm³		

The uncertainty can be determined for each of the measuring instruments — pipette, volumetric flask and burette.

$$\% \text{ uncertainty} = \frac{\text{instrument uncertainty}}{\text{quantity measured}} \times 100$$

The choice of indicator depends on the type of titration. See Table 7.

Table 7

Indicator	Colour in acid	Colour in alkali	Titrations suitable for
Methyl orange	Red	Yellow	Strong acid–strong base Strong acid–weak base
Phenolphthalein	Colourless	Pink	Strong acid–strong base Weak acid–strong base

Some of the solution from the burette may run down the inside of the conical flask, rather than into the solution it contains. The addition has been measured by the burette, so you must make sure that all the solution added has a chance to react. Sometimes, you can achieve this by swirling the solutions together, but deionised water can also be squirted down the side of the flask to wash everything into the solution below. Although the solution is undoubtedly diluted, the amount, in moles, present in the pipetted solution will not have been changed and it is this that is being titrated.

Tip

Make sure you remove the funnel used to fill the burette before titrating.

Tip

When carrying out a titration, keep the solution from the first accurate titration to achieve consistency by colour matching.

Tip

When calculating the uncertainty for a titre reading remember that the uncertainty applies for each reading, so 2 × instrument uncertainty is used.

Tip

When calculating a concentration from a titration, it is best to keep intermediate answers in your calculator, and use these in subsequent steps. This will minimise rounding errors in the final answer.

Knowledge check 9

What is the colour change of phenolphthalein indicator when a solution of sodium hydroxide solution of unknown concentration is titrated with sulfuric acid solution?

Worked example

In a titration, $25.0\,cm^3$ of potassium hydroxide solution is added to a conical flask using a pipette. Three drops of methyl orange indicator are added and nitric acid added from a burette until the indicator changes colour. Which of the following would lead to the titre being larger than it should be?

A Rinsing the conical flask with water before adding the potassium hydroxide solution.

B Rinsing the burette with water before filling it with hydrochloric acid.

C Rinsing the pipette with water before filling it with potassium hydroxide solution.

D Adding extra drops of indicator.

Answer

Rinsing the burette with water before use will make the hydrochloric acid in the burette more dilute, and so more of it will be needed for reaction, and so a larger titre value.

Rinsing the pipette with water would dilute the sodium hydroxide, so less hydrochloric acid would need to be added for neutralisation and a smaller titre value.

Adding extra drops of indicator would not have an effect.

Hence the answer is B.

Worked example

Describe how you would dilute $25.0\,cm^3$ of oven cleaner to $250\,cm^3$ and safely transfer $25.0\,cm^3$ of the diluted solution to a conical flask.

Answer

Rinse a pipette with oven cleaner.

Use a safety pipette filler with the pipette to draw up $25.0\,cm^3$ of oven cleaner and place into a $250\,cm^3$ volumetric flask.

Make up to the mark by adding deionised water until the bottom of the meniscus is on the mark.

Stopper the flask and invert to mix thoroughly.

Rinse a pipette with the diluted solution.

Transfer $25.0\,cm^3$ of diluted solution using the pipette and safety filler to a conical flask.

PAG3 Enthalpy determination

Determination of enthalpy change of combustion

The simple apparatus used to measure the enthalpy change of combustion of a liquid fuel is shown in Figure 9. The calorimeter should be insulated from its surroundings and usually contains water. An accurate thermometer measures temperature rise and the energy change is calculated using:

energy change = mass × specific heat capacity × temperature change

$$q = mc\Delta T$$

Enthalpy change of combustion is the enthalpy change when 1 mol of a substance is burned completely in excess oxygen with all reactants and products in their standard states under standard conditions.

The energy change is scaled to find the energy change for 1 mol of fuel.

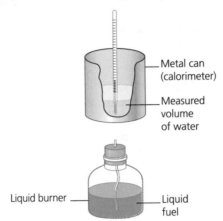

Figure 9 Apparatus used to determine the enthalpy change of combustion of a volatile liquid such as ethanol

The method to determine the enthalpy of combustion of a fuel such as ethanol or methanol is:

- Accurately measure $100\,cm^3$ of water into a calorimeter/beaker. ($100\,cm^3$ is the same as $100\,g$ water as the density of water is $1\,g\,cm^{-3}$).
- Weigh a spirit burner containing the liquid to be burnt.
- Measure the initial temperature of the water using a thermometer (T_1).
- Use the spirit burner to heat the water.
- Stop heating when there is a reasonable temperature rise (15°C). Stir and measure the final temperature (T_2) of the water using a thermometer.
- Reweigh the spirit burner.
- Calculate temperature change (ΔT) $= T_2 - T_1$ and the heat energy change using $q = mc\Delta T$.
- Calculate mass of fuel used in the burner by subtraction, and calculate the number of moles of fuel used using moles = mass/M_r.
- Calculate the energy change per mole of fuel used.

The value for the enthalpy value of combustion determined experimentally is frequently *less* exothermic than the value found in data books. Reasons for this are:

- heat losses to the surroundings from the spirit burner, wick and calorimeter (the flame is affected by draughts)
- loss of fuel from the wick or burner, by evaporation
- loss of water by evaporation
- incomplete combustion of the fuel, leaving soot on the bottom of the calorimeter, and the metal can also takes in heat
- heat used to raise temperature of calorimeter
- the reaction is unlikely to occur under standard conditions, especially temperature

Simple experiments can be *improved* to obtain more accurate values by
- using a draught shield to reduce heat loss to the surroundings
- using a lid on the calorimeter to reduce heat loss to the surroundings
- minimising the distance between the flame and the calorimeter
- insulating the calorimeter and the spirit burner to reduce heat loss
- using a top on the spirit burner with wick protruding to minimise evaporation
- if possible burn in a supply of pure oxygen to prevent incomplete combustion

Determination of enthalpy change of neutralisation

Enthalpy changes in solution, such as **enthalpy change of neutralisation** can be measured using insulated plastic cups as calorimeters as shown in Figure 10.

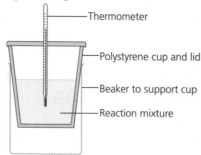

Figure 10 Apparatus used to determine enthalpy
change of neutralisation

The method to determine the enthalpy change of neutralisation is:
- Place a polystyrene cup in a glass beaker for support.
- Rinse a measuring cylinder with $1.0\,mol\,dm^{-3}$ HCl and then measure $25\,cm^3$ of this acid and transfer into the polystyrene cup.
- Stir the acid with a thermometer and record the temperature.
- Rinse a second measuring cylinder with $1.0\,mol\,dm^{-3}$ NaOH(aq) and then measure out $25\,cm^3$ of NaOH(aq).
- Add the NaOH(aq) to the acid, stir and record the highest temperature reached.
- Calculate temperature change $\Delta T = T_2 - T_1$ and the heat energy change in joules using $q = mc\Delta T$.
- Calculate the amount in mol of acid used, the amount in mol of water formed and the enthalpy of neutralisation.

This method involves measuring the temperature before mixing and the maximum temperature after mixing the reactants and finding the difference.

An alternative method is to record the temperature against time, and plot a graph, extrapolating the cooling curve after the reaction takes place in order to calculate the temperature change (ΔT) at the point of mixing the reactants, as shown in Figure 11. The extrapolation method is more important when the reaction is highly exothermic because more heat energy is lost at the point of reaction compared with that of a less exothermic reaction. Hence simply measuring the temperature difference may significantly underestimate the temperature change and the value of ΔH.

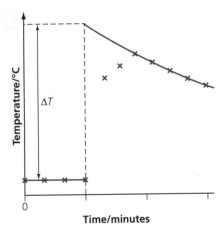

Figure 11 Estimating the maximum temperature of a neutralisation reaction

Worked example

In an experiment to determine the enthalpy change of combustion of butan-1-ol (C_4H_9OH) the following results were obtained:

mass of burner + butan-1-ol = 55.40 g

mass of burner + butan-1-ol after burning = 53.89 g

initial temperature of water = 20.2°C

final temperature of water = 44.4°C

mass of water in calorimeter = 100 g

a What is the enthalpy change of combustion?

b In this experiment the student used a thermometer with uncertainty ±0.1°C in each reading. Calculate the percentage uncertainty in the temperature rise.

Answers

a mass of butan-1-ol burned = 55.40 − 53.89 = 1.510 g

M_r butan-1-ol = 74.0

moles of butan-1-ol = $\dfrac{\text{mass}}{M_r}$ = $\dfrac{1.510}{74.0}$ = 0.0204 mol

energy absorbed by water = $q = mc\Delta T$ = 100 × 4.18 × 24.2 = 10 116 J

energy change of combustion = $\dfrac{10\,116}{0.0204}$ = −496 kJ mol⁻¹

b % uncertainty = $\dfrac{\text{instrument uncertainty}}{\text{quantity measured}}$ × 100

$\dfrac{2 \times 0.1}{24.2}$ × 100 = 0.8% (to 1 s.f.)

Tip

In this calculation the answer **a** is given to three significant figures, as this is the number of significant figures of the least accurate measurement (temperature, mass of water).

Tip

When two readings are taken, the uncertainty must be multiplied by 2.

PAG4 Qualitative analysis of ions

Knowledge check 15

Describe how you would prove that a sample of a fertiliser contained ammonium ions.

Qualitative testing for ions is usually carried out on a test tube scale. Many of these tests rely on the formation of a precipitate (ppt). A precipitate is a solid formed when two ionic solutions are mixed. It is formed because one of the combinations of ions from the solutions gives an insoluble compound. For example, when barium chloride solution is added to sodium sulfate solution, the barium ions and the sulfate ions react and a precipitate of insoluble barium sulfate forms.

$$Ba^{2+}(aq) + SO_4^{2-}(aq) \rightarrow BaSO_4(s)$$
from barium chloride ... from sodium sulfate

The methods for various ion tests are shown in Table 8.

Table 8

Ion	Experimental method	Result
Cation tests		
Ammonium ion (NH_4^+)	*Warm* the sample with sodium hydroxide solution in a test tube and test the gas produced with *moist* red litmus/universal indicator paper or test the gas produced with a glass rod dipped in concentrated HCl(aq)	Pungent-smelling gas changes moist red litmus paper/universal indicator paper to blue. The gas is alkaline (ammonia) White fumes of ammonium chloride, indicate ammonia produced This indicates that the ion in the salt was ammonium
Anion tests		
Chloride (Cl^-)	Make a solution of the compound and acidify using dilute nitric acid solution Add some silver nitrate solution followed by aqueous ammonia	White ppt which dissolves in dilute aqueous ammonia to give a colourless solution
Bromide (Br^-)	Make a solution of the compound and acidify using dilute nitric acid solution Add some silver nitrate solution followed by aqueous ammonia	Cream ppt which is insoluble in dilute aqueous ammonia but dissolves in concentrated aqueous ammonia to give a colourless solution
Iodide (I^-)	Make a solution of the compound and acidify using dilute nitric acid solution Add some silver nitrate solution followed by ammonia solution	Yellow ppt which is insoluble in dilute and concentrated aqueous ammonia
Sulfate (SO_4^{2-})	Make a solution of the compound. Acidify with dilute hydrochloric acid. Add some barium chloride solution	White ppt
Carbonate (CO_3^{2-})	Add some dilute nitric acid and test the gas produced with limewater	Effervescence and the gas changes colourless limewater milky

Nitric acid solution is often added in the test for halide ions. It removes other ions such as carbonate ions (CO_3^{2-}) and hydroxide ions (OH^-) which may be present and would react with the silver ions in the silver nitrate solution forming precipitates that would interfere with the test.

In the test for sulfate ions, the barium chloride solution is often acidified with hydrochloric or nitric acid, which reacts with and removes any carbonate ions present. Barium carbonate is a white insoluble solid that would be indistinguishable from barium sulfate.

It is important to note that these tests can be used the other way round. For example, to identify two solutions as nitric acid or silver nitrate a carbonate solution and a chloride solution can be used, giving the results shown in Table 9.

Table 9

Solution	Add sodium carbonate solution	Add sodium chloride solution
Nitric acid	Effervescence	No reaction
Silver nitrate	White ppt	White ppt

(A-level only) You also need to know the following tests:

Test for metal ions: Add a few drops of sodium hydroxide solution or ammonia solution to a solution of the metal ion and note the colour of the precipitate formed. Add excess and note the effect, if any, on the precipitate. See Table 10.

> **Tip**
>
> When carrying out a battery of tests to determine the anion present the tests should be carried out in sequence — carbonate, sulfate and then halide. This is because barium carbonate and silver sulfate are both insoluble.

Table 10

Metal ion	Cu^{2+}	Fe^{2+}	Fe^{3+}	Mn^{2+}	Cr^{3+}	Zn^{2+} (OCR B spec AS)	Al^{3+} (OCR B spec AS)
Effect of adding NaOH (aq) until in excess	Blue ppt	Green ppt	Brown ppt	Buff/pink ppt that oxidises to brown Mn_2O_3 in air	Green ppt that dissolves in excess to form a green solution	White ppt that dissolves in excess to form a colourless solution	White ppt that dissolves in excess to form a colourless solution
Effect of adding NH_3(aq) until in excess	Blue ppt that dissolves in excess to form a dark blue solution	Green ppt	Brown ppt	Buff/white ppt that oxidises to brown Mn_2O_3 in air	Green ppt	White ppt that dissolves in excess to form a colouress solution	White ppt

(OCR B specification (AS) only) You also need to know the following tests:

Test for nitrate: Warm with a spatula of Devarda's alloy and sodium hydroxide solution and test the pungent gas produced with a glass rod dipped in conc HCl(aq). White fumes are produced. The Devarda's alloy reduces the nitrate to ammonia, which is tested for using concentrated HCl(aq).

Test for metal ions using flame tests: Dip a nichrome wire in concentrated hydrochloric acid and then into the sample. Place in a blue Bunsen flame and record the colour of the flame. See Table 11.

Table 11

Metal ion	Na^+	K^+	Li^+	Ca^{2+}	Ba^{2+}	Cu^{2+}
Flame colour	Orange/yellow	Lilac	Red	Brick red	Green	Blue-green

> **Knowledge check 16**
>
> What is observed when a solution of sulfuric acid is added to some barium nitrate solution?

> **Tip**
>
> When choosing a carbonate, chloride or bromide to use as a test solution, choose a group 1 compound, as these are soluble.

> **Knowledge check 17**
>
> When sodium hydroxide is added to copper(ıı) sulfate solution a blue precipitate is formed. Write an ionic equation for this reaction.

Skills Guidance

Worked example

Four solutions to be identified in the laboratory are:

- ammonium sulfate
- potassium sulfate
- potassium iodide
- sodium carbonate

In a practical, 1 cm^3 of each solution was added to separate test tubes and 1 cm^3 of dilute hydrochloric acid added. Effervescence occurred in one test tube and the gas was collected and bubbled into colourless limewater which changed to milky.

Describe how you would continue this experiment to positively identify all of the solutions.

Answer

The reaction described identifies the sodium carbonate solution.

A good way of planning your work is to quickly draw a table showing the solutions you use and the observations that would occur. Of the three remaining substances two are sulfates, which could be identified using barium chloride solution and then distinguished between by testing for ammonium ion. The iodide ion can be tested for by adding acidified silver nitrate solution. See Table 12.

Table 12

	Add barium chloride solution	Warm with sodium hydroxide solution	Add silver nitrate solution followed by dilute aqueous ammonia
Potassium sulfate	White ppt	No reaction	
Ammonium sulfate	White ppt	Gas released which changes indicator paper blue	
Potassium iodide	No reaction		Yellow ppt

Description of experiment

1 cm^3 of the remaining three solutions was placed in separate test tubes and some dilute hydrochloric acid added followed by some barium chloride solution. A white precipitate was formed in two of the test tubes. Fresh 1 cm^3 samples of these two solutions were warmed with dilute sodium hydroxide solution.

In one test tube a gas was produced and was tested with moist universal indicator paper that turned blue, indicating an alkaline gas produced and that this solution was ammonium sulfate. The other test tube was the potassium sulfate solution.

To the final unidentified sample silver nitrate solution was added, followed by dilute aqueous ammonia and a yellow ppt insoluble in dilute aqueous ammonia was produced, identifying the solution as potassium iodide.

Worked example

Another style of question is to complete deductions from given observations. Look carefully at the test column, and note the reagent added. This should aid you in making a deduction. In the example shown in Table 13 the reagents have been highlighted in bold.

Table 13

Test	Observations	Deductions
1 Add a spatula measure of Y to a test tube one third full of **sodium hydroxide solution** and *warm gently* Carefully smell any gas given off and test it with moist universal indicator paper	Universal indicator turns blue Pungent/choking smell	Basic gas produced Ammonia gas An ammonium cation could be present
2 Add a spatula measure of Y to a test tube containing 1 cm³ of dilute **nitric acid** Add four drops of **barium chloride solution**	No effervescence White precipitate forms	A carbonate anion is not present A sulfate anion is present

PAG5 Synthesis of an organic liquid

The exact method to prepare an organic liquid depends on the actual liquid being prepared, but in general, an organic liquid can be synthesised and purified using several stages, as shown in Figure 12.

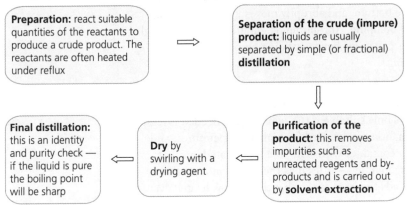

Figure 12 Synthesis of an organic liquid

Preparation

When preparing an organic liquid, the reactions are often slow, as the organic reactants contain strong covalent bonds. It is often necessary to heat the reactants under **reflux** for some time, when preparing an organic liquid.

Before reflux the reactants are often added slowly to the reaction flask, with cooling. This is because the reaction is often exothermic and adding reactants slowly with

Reflux is the continual boiling and condensing of a reaction mixture to ensure that reaction takes place without the contents of the flask boiling dry.

cooling dissipates the heat, prevents the temperature of the reaction increasing and avoids dangerous splashing or the formation of side products.

Heating under reflux involves attaching a water-cooled condenser vertically to the reaction flask, as shown in Figure 13. Vapour from the boiling reaction mixture condenses and flows back into the reaction flask. The condenser prevents vapour escaping and so the reactants can be boiled for a long period without any loss of vapour. Note that there is no stopper or thermometer at the top of the apparatus.

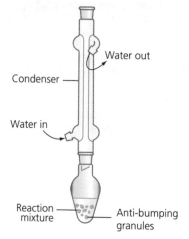

Water out

Condenser

Water in

Reaction mixture

Anti-bumping granules

Figure 13 Quickfit® apparatus for reflux. Quickfit® has ground glass joints so that a good connection is made

Tip

Note that water goes in at the bottom of the condenser. This ensures that the outer water jacket of the condenser is completely filled and an effective cooling system is set up. Remember that water enters lower than it leaves.

Anti-bumping granules are small rough pieces of silica or unglazed pottery which are added to the reaction mixture before reflux. Most liquids do not boil smoothly — often a large bubble forms at the bottom, and abruptly moves upwards causing hazardous splashing. This is called bumping. Anti-bumping granules provide a rough surface on which small gas bubbles can grow, hence avoiding bumping. *Anti-bumping granules promote smooth, even boiling.*

Most organic compounds are flammable, so often when heating a flameless method is used. Instead of a Bunsen burner a *water bath*, *sand bath* or *electric heating mantle* can be used. A water bath is only suitable if the temperature needed is less than 100°C. An electric heating mantle can be used for higher temperatures, and sometimes a sand bath — simply a container with sand in it and heated by a hot plate, or a layer of sand in a heating mantle — is used as the sand conducts the heat to the reaction flask, and spreads the heat out so that the flask is heated evenly without the need for stirring.

Separation of the crude product

After reflux, allow the apparatus to cool and rearrange for **distillation**. To do this the condenser is removed, a still head is inserted in the flask, a thermometer placed in the still head and the water-cooled condenser attached sideways. A receiver is attached to the end of the condenser. Distillation apparatus is shown in Figure 14.

Anti-bumping granules must also be used for smooth boiling in distillation, because if bumping occurs the liquid can splash over into the condenser, causing an impure product or can blow the distillation apparatus apart. The distillate is the crude product, and is collected over a boiling point range.

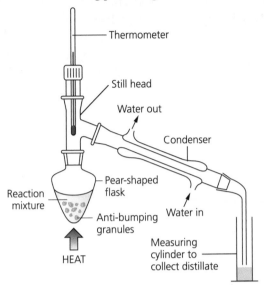

Figure 14 Distillation Quickfit® apparatus

> **Tip**
>
> Before carrying out reflux or distillation, make sure that the joints in your apparatus are secure, so that volatile chemicals do not escape during the reaction. It is a good idea to mention this in a risk assessment, and ask your teacher to check your apparatus before heat is applied.

Drawing diagrams

In exam questions you may be asked to draw diagrams of the apparatus set up for reflux or for distillation. It is important that you draw these diagrams as cross sections, as shown in Figures 15 and 16, not as three-dimensional pictures as shown in Figures 13 and 14. Diagrams must be labelled. Note that in the distillation apparatus, there is a vent on the receiver, so that the apparatus is not closed. An alternative to this is to collect the product in an open measuring cylinder, as shown in Figure 14.

> **Tip**
>
> Diagrams of distillation and reflux do not need to show Bunsen burners or clamps.

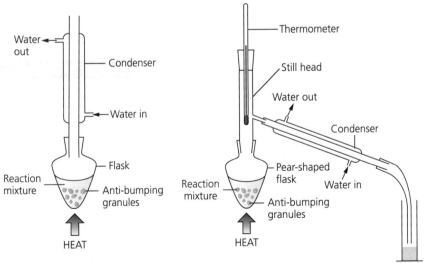

Figure 15 Diagram of reflux

Figure 16 Diagram of distillation

> **Tip**
>
> The bulb of the thermometer is opposite the exit to the condenser so that the temperature of the vapour exiting, which condenses to form the product, can be recorded.

Purification of the product

The crude (impure) liquid product is often contaminated with by-products or unreacted reactants. Purification of the liquid can be carried out using **solvent extraction** in a separating funnel (Figure 17). This involves shaking the organic liquid with an aqueous solution. The impurities are more soluble in the aqueous solution and move into it — they are extracted. The organic liquid and aqueous solution are immiscible and can be separated.

Immiscible liquids are those that do not mix, and form two layers.

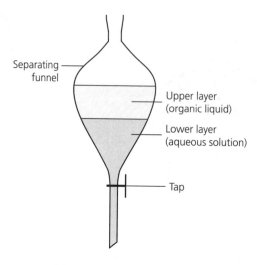

Figure 17 Separating funnel

Solvent extraction

- Place the organic liquid in a separating funnel and add a portion (e.g. 10 cm³) of aqueous solution (most often this is **sodium hydrogen carbonate solution** to remove acidic impurities in the organic liquid).
- Stopper and shake, releasing the pressure (often due to build-up of carbon dioxide in an acid + carbonate reaction) by inverting and opening the tap.
- Allow the separating funnel to stand until the layers settle and separate.
- Remove the stopper and open the tap, to run off the bottom layer into a beaker. When the bottom layer is nearly all drained, close the tap partially to slow down the flow and avoid any of the top layer leaving the funnel. Run off the second layer into a separate beaker.
- Discard the aqueous layer.
- Place the organic layer back into the separating funnel and repeat the process, using another portion of the aqueous solution. It is more effective to carry out solvent extraction twice with small portions of solvent rather than once with a large portion.

How do you decide which layer is the organic liquid?

One way to do this is to refer to the density. The more dense liquid will be the bottom layer. Alternatively, add some water and observe which layer increases in size — this will be the aqueous layer.

Tip

If a question asks about solvent extraction always check if the densities are given, and determine which is the organic layer.

Knowledge check 18

Eucalyptus oil (density 0.922 g cm⁻³) can be produced by steam distilling eucalyptus leaves, washing the distillate with sodium chloride solution (density 1.22 g cm⁻³) and separating. Draw a labelled diagram of the apparatus used to separate the oil.

Drying

The purified organic liquid may still contain water, and must be dried. The method for this is:

- Add a spatula of a drying agent, for example anhydrous calcium chloride or anhydrous magnesium sulfate, to the organic liquid in a beaker or conical flask.
- Swirl.
- Add more of the drying agent until the liquid changes from cloudy to clear.
- Filter (Figure 18) or **decant** off the liquid into a clean, dry flask.

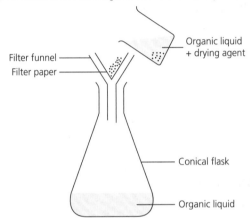

Filter funnel
Filter paper
Organic liquid + drying agent
Conical flask
Organic liquid

Figure 18 Gravity filtration

Final distillation

The product can be redistilled. The boiling point measured during the distillation indicates the *identity* and the *purity* of the liquid product. If the boiling point is sharp, then the product is pure. If it covers a range, then impurities may be present. If the boiling point recorded during distillation is the same as that given in data books, then the correct product has been obtained.

If the boiling points of the reactants and the product are very similar, often fractional distillation (Figure 19) may be used instead of simple distillation.

In fractional distillation if the flask contains a mixture of liquids, the boiling liquid in the flask produces a vapour which is richer in the most volatile of the liquids present — the one with the lowest boiling point. Most of the vapour condenses in the column and runs back. As it does so, it meets more of the rising vapour. Some of the vapour condenses and some of the liquid evaporates.

In this way, the mixture evaporates and condenses repeatedly as it rises up the column. But every time it does so, the vapour becomes richer in the most volatile liquid present. At the top of the column, the vapour can contain close to 100% of the most volatile liquid. So, during fractional distillation, the most volatile liquid with the lowest boiling point distils over first, then the liquid with the next lowest boiling point and so on. The boiling point can be measured as the liquid distils over during fractional distillation. Impurities increase the boiling point and the range over which the liquid boils.

Tip

A common error is to state 'decant off the anhydrous calcium chloride'. Be careful — at this stage the calcium chloride is no longer anhydrous, and it is the liquid that is decanted off.

Decanting means to carefully pour a liquid from one container to another, in order to leave any solid in the bottom of the original container.

Knowledge check 19

Name a chemical used to dry eucalyptus oil.

Tip

If a liquid is pure it should distil over a narrow range, at the expected boiling point.

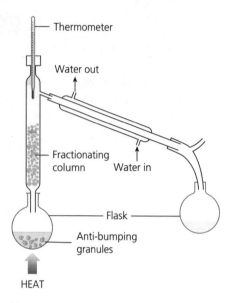

Figure 19 Fractional distillation diagram

Why is the percentage yield less than 100%?

There are theoretical and practical reasons for this.

Theoretical reasons:

- Side reactions occur, so by-products may be produced instead of the expected product.
- The reagents used may be impure.
- The reaction is incomplete.

Practical reasons:

- Some product is lost in purification steps, for example in washing, separating in separating funnel, and in transfer between apparatus.
- Some product is lost in distillation.

Every organic liquid will be prepared in a slightly different method. It is important that you understand the general preparation methods listed and apply them to different situations, as shown in the worked examples.

Knowledge check 20

A common step in purification of organic liquids is shaking with water in a separating funnel. Why is this not an appropriate method for purifying ethanoic acid?

Worked example

The ester ethyl ethanoate was prepared from ethanol and ethanoic acid, using the method given below:

- Mix 12.0 cm^3 of ethanoic acid (density 1.05 g cm^{-3}) and 10.0 cm^3 of ethanol (density 0.789 cm^{-3}) in a pear-shaped flask.
- Add 10 cm^3 of concentrated sulfuric acid slowly with cooling and shaking.
- Add some anti-bumping granules and heat under reflux for 20 minutes.
- Distil off the ethyl ethanoate and collect the fraction between 74 and 79°C.
- Place the crude ethyl ethanoate in a separating funnel and shake with sodium carbonate solution. Invert the funnel and open the tap occasionally.

→

- Allow the layers to separate and discard the lower aqueous layer.
- Add some calcium chloride solution to the ethyl ethanoate, to remove any ethanol impurities, shake, and run off the aqueous layer.
- Add a spatula of anhydrous calcium chloride and shake. Repeat until the ester is clear.
- Decant the liquid into a clean, dry, pear-shaped flask and redistill, collecting the ethyl ethanoate (density $0.92\,\mathrm{g\,cm^{-3}}$) at the boiling point.

a Why must the concentrated sulfuric acid be added slowly and with cooling?
b Why is concentrated sulfuric acid used in this reaction?
c Write a balanced symbol equation for the reaction to prepare ethyl ethanoate.
d What are the nature and purpose of anti-bumping granules?
e How is the apparatus set up to heat under reflux?
f What is the function of the sodium carbonate solution?
g Why is the tap opened from time to time?
h Is the aqueous layer the top or bottom layer?
i Why is anhydrous calcium chloride added until the ester is clear?
j Why must the pear-shaped flask be clean and dry?
k Calculate the percentage yield if $8.2\,\mathrm{cm^3}$ of ethyl ethanoate (density $0.92\,\mathrm{g\,cm^{-3}}$) is collected.

Answers

a The concentrated sulfuric acid on dilution gives out a lot of heat. The slow addition with cooling dissipates the heat and avoids splashing which would occur if the mixture gets hot.
b It is a catalyst for the reaction. It is also a dehydrating agent and removes water, promoting the forward reaction.
c $CH_3CH_2OH + CH_3COOH \rightleftharpoons CH_3COOCH_2CH_3 + H_2O$
d Anti-bumping granules are small pieces of silica or broken unglazed pottery. They promote smooth even boiling.
e A condenser is fitted vertically to the pear-shaped flask.
f The distillate contains traces of unreacted ethanoic acid and concentrated sulfuric acid. The sodium carbonate solution removes these. (Sometimes the distillate is then washed with water, to remove traces of sodium carbonate solution.)
g The neutralisation of acid with sodium carbonate produces carbon dioxide gas; opening the tap releases this and avoids a build-up of pressure that might blow the stopper out of the funnel.
h The density of water and aqueous solutions is $1.0\,\mathrm{g\,cm^{-3}}$. The density of the ester is $0.92\,\mathrm{g\,cm^{-3}}$. Hence the aqueous layer is the bottom, denser layer.
i To remove water and dry the ester.
j The ester has just been purified and dried.
k
$$\text{density of ethanol} = \frac{\text{mass}}{\text{volume}} \qquad \text{density of ethanoic acid} = \frac{\text{mass}}{\text{volume}}$$

$$0.789 = \frac{\text{mass}}{10.0} \qquad\qquad\qquad 1.05 = \frac{\text{mass}}{12.0}$$

$$\text{mass} = 7.89\,\mathrm{g} \qquad\qquad\qquad \text{mass} = 12.6\,\mathrm{g}$$

$$\text{amount in moles} = \frac{\text{mass}}{M_r}$$

$$= \frac{7.89}{46.0}$$

$$= 0.17$$

$$\text{amount in moles} = \frac{\text{mass}}{M_r}$$

$$= \frac{12.6}{60.0}$$

$$= 0.21$$

The ethanoic acid is in excess.

$$\text{moles of ethyl ethanoate} = 0.17 = \frac{\text{mass}}{M_r} = \frac{\text{mass}}{88.0}$$

$$\text{moles of ethyl ethanoate} = 0.17 \times 88.0 = 14.96\,\text{g}$$

$$\text{density of ethyl ethanoate} = 0.92 = \frac{14.96}{\text{volume}}$$

$$\text{volume} = \frac{14.96}{0.92} = 16.26\,\text{cm}^3$$

$$\text{\% yield} = \frac{\text{actual yield}}{\text{theoretical yield}} \times 100$$

$$= \frac{8.2}{16.26} \times 100 = 50.4\%$$

> **Knowledge check 21**
>
> Name a technique that could be used to boil an organic reaction mixture for 30 minutes.

Worked example

To prepare ethanal (b.p. 21°C) in the laboratory, the following method was used.

7.5 g of sodium dichromate was placed into a distillation flask with 15 cm³ of water. The apparatus was set up for distillation and a mixture of 3 cm³ of concentrated sulfuric acid and 6 cm³ of ethanol (density = 0.79 g cm⁻³) placed into a tap funnel, and slowly added to the distillation flask. The flask was occasionally heated very gently until about 6 cm³ of distillate was collected in a receiving flask surrounded by a mixture of ice and water.

a Identify two aspects of this preparation that might be hazardous. What steps would you take to minimise risks from these hazards?

b When the reaction is over, the remaining colour in the flask is orange. Explain which reactant is in excess.

c Why is distillation used instead of reflux in this experiment?

d Suggest why the distillate is best collected in a flask surrounded by ice.

e 2 cm³ of the distillate containing ethanal, ethanol and water is added to a beaker of boiling water.

  i Explain what would happen.

  ii Describe how you would obtain a sample of ethanal from a mixture of ethanal, ethanol and water. Include in your answer a description of the apparatus you would use and how you would minimise the loss of ethanal. Your description of the apparatus can be either a description in words or a labelled sketch.

f What is observed if the distillate is warmed with Tollens' reagent?

→

Answers

a The concentrated sulfuric acid is corrosive. It must be handled with care, and the handler should wear safety goggles and gloves.

Ethanol is flammable. Keep it away from naked flames when measuring out. Consider using an electric heating mantle.

Use anti-bumping granules to prevent dangerous splashing and bumping.

b The orange colour indicates that the sodium dichromate is in excess. If it was all reduced, then a green colour would be present.

c Distillation means that the ethanal is distilled off immediately and separated from the oxidising agent. Hence further oxidation is prevented. Reflux would mean that the ethanal would still be in contact with the oxidising agent and would fully oxidise to ethanoic acid.

d Ethanal has a low boiling point of 21°C, close to room temperature. The ice cools the flask and ensures that ethanal is collected as a liquid, and does not evaporate.

e i Ethanal and ethanol can hydrogen bond with water — they are miscible with water. The liquids would boil as both have boiling points less than 100°C.

ii To separate these liquids the mixture could be distilled by placing the mixture in a flask, attaching a still head with thermometer and a water-cooled condenser sideways. An ice-water-cooled receiving flask (to reduce evaporation of ethanal) should be used to collect the ethanal at 21°C. (Note the temperature then rises and a different flask could be used to collect the ethanol at 76°C.) An alternative answer would be a diagram similar to that in Figure 16, with the receiving flask cooled in ice water.

f A silver mirror as the ethanal is oxidised.

PAG6 Synthesis of an organic solid

The exact method to prepare an organic solid depends on the actual solid being prepared, but in general, an organic solid can be synthesised and purified using several stages as shown in Figure 20.

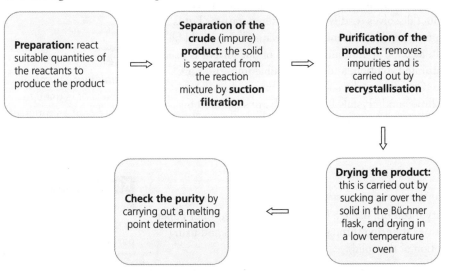

Figure 20 Synthesis of an organic solid

Tip

Remember that acidified potassium or sodium dichromate are oxidising agents and change colour from orange to green when warmed with a reducing agent such as ethanol.

Tip

When drawing a distillation diagram make sure the thermometer bulb is level with the outlet, the water direction is correct in the condenser, and it is an open system.

Preparation and separation of the crude product

To prepare an organic solid, solutions of the reactants are often added together at room temperature and the product precipitates out. Alternatively the reactants are refluxed together, the anti-bumping granules decanted off and the solid forms on cooling and crystallising.

Suction filtration (filtration under reduced pressure) is used to separate the product (Figure 21). It is faster than normal filtration and the solid is left quite dry.

To set up the apparatus for suction filtration, place a circle of filter paper into a Büchner funnel and place in a stopper in a Büchner flask. Connect the Büchner flask to a suction pump and pour the mixture into the funnel. The suction draws the liquid through into the Büchner flask and leaves the crude product in the filter paper.

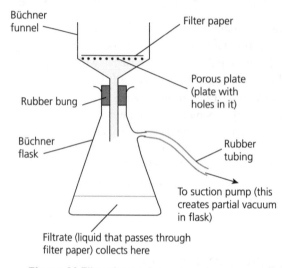

Figure 21 Filtration under reduced pressure

Purification and drying

To obtain a pure solid product, recrystallisation is usually required. A solvent is chosen in which the desired product dissolves readily at higher temperatures but is only slightly soluble at room temperature. The solvent is often water, but it could be another suitable liquid, such as an alcohol. A minimum volume of hot solvent is used to dissolve the solid, making a saturated solution, and any insoluble impurities are then removed by gravity filtration. On cooling, the solubility of the product drops, causing it to recrystallise from solution and crystals form, which are separated by filtration under reduced pressure. Impurities remain dissolved in the solution. Slow crystallisation is preferred, as fast crystallisation can cause some soluble impurities to become trapped in the crystals.

The method of recrystallisation is:

- Dissolve the impure crystals in the *minimum volume of hot solvent*.
- Filter the hot solution by gravity filtration, using a hot funnel and fluted filter paper, to remove any insoluble impurities (filtering through a hot filter funnel and using fluted paper prevents precipitation of the solid).

Tip

A *minimum volume of hot solvent* is used to ensure that as much of the solute is obtained as possible.

Knowledge check 22

Name the method used to purify some solid paracetamol.

- Allow the solution to cool and crystallise (the impurities will remain in solution). Sometimes scratching the side of the flask with a glass rod or adding a small seed crystal can aid crystal formation.
- Filter off the crystals using suction filtration.
- Wash by pouring over some *ice-cold solvent* which removes any aqueous impurities. (The solvent is cold to prevent the crystals from dissolving.)
- Dry by sucking air over the crystals in the Büchner flask and then in a low-temperature oven. Alternative methods of drying include placing in a **desiccator** (Figure 22) with a drying agent.

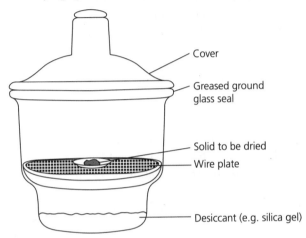

Figure 22 A desiccator

Further recrystallisation can be carried out to obtain a purer product.

Checking the purity of an organic solid

The melting point of a substance is not the exact point at which it melts, but rather the range of temperatures from when the sample starts to melt until it has completely melted. The greater the range, the more impurities are present. A pure substance melts over a narrow range of temperatures, while an impure substance melts over a wide range of temperatures and at a lower temperature than the pure substance. The melting points of almost all substances are available in data tables.

To check the purity of a solid, a melting point can be determined using the following method:

- Place some of the solid in a melting point tube *sealed* at one end.
- Place in melting point apparatus and heat slowly.
- Record the temperature at which the solid starts to melt and the temperature at which it finishes melting.
- Repeat and average the temperatures.
- Compare the melting point with known values in a data book.

Tip

A melting point range of less than 2°C indicates a fairly pure substance.

Knowledge check 23

Suggest how you would know that a sample of aspirin was impure, from the melting point determination.

Tip

The *identity* of an organic substance can be confirmed by taking a melting point of a pure sample and comparing it with known values in a data book.

Using chromatography

Chromatography can be used to determine that the solid has been synthesised, but it does not give information about purity unless a chromatogram of the prepared solid is compared with a chromatogram sample of pure solid. A thin layer plate may be used, or paper chromatography, but the method is similar.

- Draw a pencil line 1.5 cm from the bottom of the thin layer chromatography plate and place two pencil crosses on the line.
- Place a drop of the purified solid on a watch glass and dissolve in a few drops of solvent such as ethanol. Use a capillary tube to place a spot of the solvent on a pencil cross. Allow the spot to dry and repeat 3–4 times, ensuring the diameter of the spot is no more than 0.5 cm. This produces a concentrated spot. Repeat this for the pure solid.
- Place solvent in a beaker to a depth of 1 cm.
- Place the chromatography plate in the beaker and cover with a lid.
- Allow the solvent to run up the chromatography plate for about 30 minutes. When it has almost reached the top of the plate, remove from the beaker and mark the line of the solvent front with a pencil.
- Place the plate in a fume cupboard until all of the solvent has evaporated and the plate is dry. If the spots are colourless then there are different ways to view them.
  - Place the plate under an ultraviolet lamp and mark the locations of the substances using a pencil.
  - In a fume cupboard place the plate in a beaker containing iodine crystals and cover with a watch glass. The iodine is a locating agent which causes the spots to become brown and visible.
  - Spray with ninhydrin developing agent in a fume cupboard.
- Calculate the R_f value of each substance visible on the plate and compare to data books.

Tip

$$R_f = \frac{\text{distance moved by spot}}{\text{distance moved by solvent}}$$

Always measure to the centre of each spot.

The R_f of spot X in Figure 23, for example, is $\frac{3}{6} = 0.5$

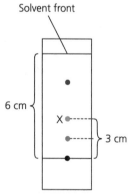

Figure 23 Determing R_f from a chromatogram

Tip

TLC can be used to monitor the course of a reaction by taking samples from the reaction mixture at regular intervals, spot on a TLC plate, and run alongside controls of the organic reactant and product.

Tip

Do not look directly into the ultraviolet light.

The chromatogram of the purified prepared solid should have similar R_f values to that of the pure solid, showing that the synthesis is successful. The chromatogram of the prepared sample may in addition contain spots with R_f values similar to some of the reactants or to some impurities, showing that it is not completely pure. The pure sample would not have these spots. The intensity of the spot shows the amount of the substance present.

Worked example

a Why is it necessary to wear plastic gloves when holding a TLC plate?

b Why is it necessary to draw a *pencil* base line 1.5 cm from the bottom of the plate?

c Explain why the developing tank solvent is at a depth of only 1 cm.

d Explain why the developing tank is sealed with a lid when the TLC plate is placed in it.

e Explain why the TLC plate is allowed to dry in a fume cupboard.

f The substance used to develop the spots on the chromatogram is ninhydrin. State one safety precaution specific to the use of this substance, apart from the use of goggles.

Answers

a To prevent contamination of the plate by amino acids from the skin, which would interfere with results.

b The pencil line is insoluble and will not move with the solvent or interfere with results.

c If the solvent is too deep, it will dissolve away the mixture.

d If the tank is open, solvent may evaporate and not advance up the plate. Having a sealed environment also allows solvent to saturate the atmosphere inside.

e The solvent can be toxic or flammable.

f Avoid breathing vapour, wear gloves, carry out procedure in a fume cupboard.

Percentage yield

Why is the percentage yield less than 100% when preparing an organic solid?

The theoretical reasons are similar as for an organic liquid. The practical reasons why some solid is lost are:

- Some product is lost in purification steps. For example, in recrystallisation some solid may still be dissolved in the solvent.

- Some product is lost in transferring between vessels — to minimise this loss sometimes rinsing is useful.

Every organic solid will be prepared by a slightly different method. It is important that you understand the general preparation methods listed and apply them to different situations, as shown in the worked examples.

Skills Guidance

The following method can be used to prepare aspirin in the laboratory:

- Place 20.0 g of 2-hydroxybenzoic acid in a pear-shaped flask and add 40 cm³ of ethanoic anhydride ($(CH_3CO)_2O$).
- Safely add 5 cm³ of concentrated phosphoric(v) acid and heat under reflux for 30 minutes.
- Add water to hydrolyse any unreacted ethanoic anhydride to ethanoic acid, and pour the mixture into 400 g of crushed ice in a beaker.
- The product is removed by filtration under reduced pressure, recrystallised, washed with ice-cold water and dried in a desiccator.
- The melting point is then determined.

The reaction can be represented as follows.

$$HOOCC_6H_4OH + (CH_3CO)_2O \rightarrow HOOCC_6H_4OCOCH_3 + CH_3COOH$$

a Suggest the role of concentrated phosphoric(v) acid in this preparation.
b Explain how the concentrated phosphoric(v) acid is added safely.
c What is used to heat the mixture under reflux?
d What happens to the ethanoic acid formed by hydrolysis of ethanoic anhydride?
e Assuming a 70% yield, calculate the mass of 2-hydroxybenzoic acid required to form 5.0 g of pure aspirin.
f Why is the mixture poured onto crushed ice?
g Describe the properties required of a suitable solvent for use in recrystallisation.
h State what happens to insoluble and soluble impurities during recrystallisation.
i Samples of 2-hydroxybenzoic acid, pure aspirin, the crude product and the recrystallised product from this experiment were spotted onto a chromatography plate. Describe how the plate is analysed.

Answers

a It is a catalyst.
b It is added slowly with stirring to safely distribute any heat produced.
c To safely heat under reflux use a boiling water bath/sand bath/electric heater.
d Ethanoic acid is miscible with water, so it will be in the filtrate.
e moles of aspirin = 5.0/180.0 = 0.0278

There is a 70% yield, so $\frac{0.0278}{70} \times 100 = 0.040$

number of moles 2-hydroxybenzoic acid = 0.040
mass of 2-hydroxybenzoic acid = 0.040 × 138.0 = 5.52 g
f To form crystals
g A solvent is chosen in which the aspirin is more soluble at higher temperatures and less soluble at lower temperatures.
h Insoluble impurities are filtered out and soluble impurities remain dissolved in the solvent.
i R_f values are taken. The crude product contains a spot which is 2-hydroxybenzoic acid and a spot which is aspirin. The recrystallised product contains aspirin and a less intense spot due to unreacted 2-hydroxybenzoic acid as there is less impurity present. The pure aspirin only contains aspirin.

PAG7 Qualitative analysis of organic functional groups

Most organic functional groups have a characteristic qualitative test. For example, an alkene can be identified because it reacts with orange-brown bromine solution to produce a colourless solution. Functional group tests are usually carried out on a test tube scale.

Table 14 summarises some of the qualitative tests which can be used to identify the presence of functional groups in organic compounds.

Table 14

Test	Observation	Deduction
Add universal indicator paper	Changes to red	Acidic — could be a carboxylic acid or phenol
Shake with bromine water	Orange-brown solution changes to a colourless solution	C=C present (alkene) (Note that phenol decolourises bromine and a white ppt forms.)
Warm with acidified potassium dichromate(VI)	Orange solution changes to green solution	Primary/secondary alcohol or aldehyde
	Solution remains orange	Could be a tertiary alcohol (it is not oxidised)
Warm with Tollens' reagent (ammoniacal silver nitrate)	Silver mirror	Aldehyde
	Solution remains colourless	Ketone
Warm with Fehling's solution **(spec B only)**	Red ppt	Aldehyde
	Solution remains blue	Ketone
Warm with silver nitrate solution in ethanol	White ppt	Chloroalkane
	Cream ppt	Bromoalkane
	Yellow ppt	Iodoalkane
Add sodium carbonate	Effervescence — the gas produced changes colourless limewater to cloudy	Carboxylic acid — the gas produced is carbon dioxide (Note that phenol is weakly acidic and can react with bases and metals but it is not sufficiently acidic to react with carbonates.)
Add magnesium	Effervescence — a pop is heard when a lighted splint is applied to the gas produced	Carboxylic acid — the gas produced is hydrogen.
Add neutral iron(III) chloride solution **(spec B only)**	Pale yellow solution turns purple	Phenol
Add 2,4-dinitrophenylhydrazine solution	An orange ppt is formed	An aldehyde or ketone
Warm with ethanol and a few drops of concentrated sulfuric acid. Cautiously smell	Sweet smell	Ester produced, the organic substance could be a carboxylic acid.

Tip

Organic compounds are often flammable and when carrying out analytical tests it is best to heat in a water bath, or use an electric heater.

Tip

Primary alcohols and aldehydes are oxidised to carboxylic acids by acidified potassium dichromate(VI). Secondary alcohols are oxidised to ketones by acidified potassium dichromate(VI).

Tip

The lack of a hydrogen atom bonded to the carbon with –OH bonded to it makes tertiary alcohols resistant to oxidation by acidified dichromate(VI).

Worked example

The following tests were carried out on a compound A and the observations recorded in Table 15.

Table 15

	Test	Observation
1	Add a spatula of sodium carbonate to A	No bubbles produced
2	Add a few drops of A to a solution of 2,4-dintrophenylhydrazine	Orange solid forms
3	Warm a few drops of A with Tollens' reagent	The solution remains colourless

Make deductions based on these observations. What type of organic chemical is A?

Answer

In test 1, A does not react to produce bubbles of carbon dioxide so A is not a carboxylic acid. Test 2 shows that A has a carbonyl group, so it could be an aldehyde or a ketone. Test 3 shows that A is not an aldehyde, and is therefore a ketone, since it has a carbonyl group.

Worked example

A student warms some propan-1-ol with acidified potassium dichromate(VI) using the apparatus shown in Figure 24.

a State one reason why this apparatus is not suitable to use in this experiment.

b Describe a more suitable way of carrying out this test.

c Name the organic compound produced.

d What is observed in the test tube?

e If this reaction was carried out by heating for an extended time to produce a final product, suggest a suitable way of carrying out the reaction.

Answers

a Propan-1-ol is flammable.

b Warm the boiling tube gently in a water bath.

c Propanoic acid — if the alcohol is not fully oxidised, then there may also be some propanal.

d Orange solution changes to green solution.

e Reflux with some anti-bumping granules added to the mixture.

Figure 24 Heating propan-1-ol with acidified potassium dichromate(VI)

Knowledge check 25

Why does bromine water decolourise when added to an alkene?

Knowledge check 26

A silver mirror is formed when Tollens' reagent is warmed with ethanol. What metal ion is reduced in this reaction? Write an ionic equation.

Knowledge check 27

Write balanced symbol equations for the reaction of propanoic acid with magnesium and with sodium carbonate, and give observations.

Worked example

An organic compound gives an orange crystalline product with
2,4-dinitrophenylhydrazine, but does not give a silver mirror with Tollens' reagent.
Which one of the following molecules is the organic compound?

A ethyl propanoate

B methanal

C propanone

D propanoic acid

Answer

All organic structures that have carbonyl groups will give a ppt with
2,4-dinitrophenylhydrazine. Both B and C have carbonyl groups. All aldehydes give
silver mirrors with Tollens' reagent, so the structure is not B.

The answer is C.

PAG8 Electrochemical cells

Electrochemical cells produce an electric potential difference (voltage) from a redox
reaction. In an electrochemical cell the two half-reactions occur in separate half-
cells. The electrons flow from one cell to the other through a wire connecting the
electrodes.

The potential difference between the two half-cells is *maximum* when no current is
flowing and is called the cell potential (or electromotive force). The cell potential can
be measured using a high-resistance voltmeter. As very little current is drawn by the
voltmeter, each electrode is effectively in equilibrium. The measured cell potential
will be close to the standard cell potential (assuming standard conditions are used).

In the laboratory many different cells can be set up using strips of metals dipping into
solutions of their own ions connected by a high-resistance voltmeter and a salt bridge.
If the half-cell is for a system that contains two ions (e.g. Fe^{2+} and Fe^{3+}), a carbon or
platinum electrode is used.

What is a salt bridge?

The electric circuit is completed by a salt bridge connecting the two solutions. It
allows ions to flow while preventing the solutions from mixing. The charge of all ions
(cations and anions) in both half-cells must remain zero to keep the electron flow
continuing, and the salt bridge allows ions to move between half-cells and keep the
charge in each container zero.

Often a salt bridge is a piece of filter paper soaked in potassium nitrate solution.
Potassium nitrate is used because all potassium and all nitrate salts are soluble, so the
potassium nitrate does not react to produce precipitates with any of the ions in the
half-cells.

Tip

Standard conditions
for electrochemical
measurements are:
temperature of 298K,
gases at pressure
of 100 kPa, solutions
at a concentration of
$1.0\,mol\,dm^{-3}$.

Tip

Cell potentials (emf)
are at a maximum
when no current flows
because under these
conditions no energy is
lost due to the internal
resistance of the cell as
the current flows.

Knowledge check 28

State the conditions
required to measure
standard electrode
potentials.

Worked example

In an experiment the cell in Figure 25 was set up with copper as the negative electrode, and the cell potential altered by changing the positions of equilibrium in the half-cell by changing the concentration of the iron(II) ions in the solution and recording the result.

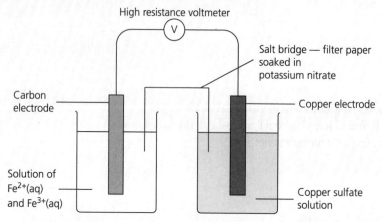

Figure 25 Measuring electrode potentials

The results for five different solutions labelled 1–5, each containing a different ratio of iron(II) ions to iron(III) ions, are shown in Table 16.

Table 16

Solution	Ratio $Fe^{2+}:Fe^{3+}$	Voltage/V
1	1:5	0.574
2	1:3	0.562
3	1:1	0.534
4	3:1	0.506
5	5:1	0.495

The electrode potentials of the two half-cells were:

$$Cu^{2+} + 2e^- \rightleftharpoons Cu(s) \qquad E^\ominus = +0.34\,V$$

$$Fe^{3+} + e^- \rightleftharpoons Fe^{2+}(aq) \qquad E^\ominus = +0.77\,V$$

a Outline a method which could be used to identify the positive electrode in an electrochemical cell.

b Why does the iron half-cell use a carbon electrode?

c Write the overall equation for the redox reaction which occurs.

d State the direction of electron flow in the external circuit.

e Calculate the electrode potential of the cell.

f Describe the function of the salt bridge.

g Explain why a piece of wire is not used instead of a salt bridge.

h State the trend in concentration of iron(III) ions in the mixture from solution 1 to solution 5.

i Using Le Chatelier's principle, explain the effect of the trend in question h on the position of equilibrium for the iron half-cell reaction and on the electrode potential.

Knowledge check 29

Write two ionic half-equations and the overall balanced equation for each of the following redox reactions. In each example, state which atom, ion or molecule is oxidised and which is reduced:

a magnesium metal with copper(II) sulfate solution

b aqueous chlorine with a solution of potassium bromide

Answers

a When the cell is set up, if the reading on the voltmeter is positive, then the metal connected to the positive terminal on the voltmeter is the positive electrode. If the reading is negative, then the metal connected to the negative terminal is the positive electrode.

b The iron half-cell uses a graphite electrode, as both iron species involved in the redox reaction are ionic and in solution. The graphite is a good conductor, has low reactivity and so does not react with the solutions.

c In the copper half-cell the reaction that occurs is:

$$Cu(s) \rightarrow Cu^{2+} + 2e^- \qquad \text{(oxidation)}$$

In the iron half-cell, the reaction that occurs is:

$$Fe^{3+} + e^- \rightarrow Fe^{2+}(aq) \qquad \text{(reduction)}$$

Hence the overall equation for the reaction occurring is:

$$Cu(s) + 2Fe^{3+}(aq) \rightarrow Cu^{2+}(aq) + 2Fe^{2+}(aq)$$

d The copper atoms form ions and give up electrons which flow from the copper half-cell through the external circuit to the iron half-cell. This is from right to left in the diagram.

e $$Cu^{2+} + 2e^- \rightleftharpoons Cu(s) \qquad E^\ominus = +0.34\,V$$

$$Fe^{3+} + e^- \rightleftharpoons Fe^{2+}(aq) \qquad E^\ominus = +0.77\,V$$

$$E^\ominus_{cell} = +0.77 - 0.34 = +0.43\,V$$

f The salt bridge is needed to complete the electrical circuit of the electrochemical cell, while keeping the solutions separated. As the concentration of copper ions increases, nitrate ions enter the solution from the salt bridge to balance the charge. As the concentration of iron(III) ions decreases, potassium ions enter the solution from the salt bridge to balance the charge.

g In a wire the conducting species are electrons, but in the half-cell the conducting species are ions. As a result, no current could flow between the half-cells using a wire. Hence the circuit would not be complete and the cell potential could not be measured.

h Moving from solution 1 to solution 5 the concentration of iron(III) ions decreases.

i To oppose the decrease in iron(III) concentration the position of the equilibrium shifts to the left-hand side of the equation:

$$Fe^{3+} + e^- \rightleftharpoons Fe^{2+}$$

This decrease in iron(III) ion solution concentration results in less positive voltages. The cell potential is calculated as:

$$E_{cell} = E(\text{reduction reaction}) - E(\text{oxidation reaction})$$

Because the electrode potential of the iron reaction becomes more negative with decreasing concentration, the difference between the electrode potential of the iron half-cell and the copper half-cell gets smaller and so the voltage reading gets smaller.

> **Tip**
>
> In a cell oxidation occurs in one half-cell and atoms are oxidised to ions as they give up electrons which build up on the metal electrode. It is the negative electrode. The electrons flow out of this electrode through the circuit towards the positive electrode.

> **Tip**
>
> At the positive electrode, ions take electrons from the metal electrode and form atoms. The electrode is positive as it loses electrons.

> **Tip**
>
> The further to the left the position of equilibrium is, the less positive (more negative) the value of the electrode potential.

PAG9 Rates of reaction: continuous monitoring method

To investigate the rate of a reaction, the method of continuous monitoring can be used. In this method, the reaction is monitored throughout its course and the amount of reactant or product is established at different time intervals throughout the reaction. There are a variety of methods that can be used to follow the progress of the reaction against time: most measure a change in the amount or concentration of a reactant or product during the reaction.

To decide on a method to monitor rate of reaction, examine the equation of the reaction and identify a reactant or a product that can be measured. Examples include:

- a gaseous product that can be monitored by measuring *volume of gas produced* or by *loss in mass* in the reaction system
- a coloured reactant or product that can be monitored using *colorimetry*
- a titratable reactant or product that can be monitored by *sampling*, *quenching* and *titrating*
- a directly measurable reactant or product, i.e. H^+ ions or OH^- ions by measuring pH using a *pH meter*

Measuring gas volume

If a gas is soluble in water, for example carbon dioxide, then a gas syringe may be attached to a sealed reaction vessel to measure the volume of gas produced (Figure 6, p. 15). Alternatively, for gases that are not very soluble in water such as oxygen or hydrogen, the gas may be collected under water in an inverted measuring cylinder (Figure 5, p. 14) or burette.

Many experiments involving the production of a gas involve heating a solid, for example decomposition of a carbonate. If a solid is heated to generate a gas, a test tube or boiling tube should be used, as shown in Figure 26, rather than a conical flask. Remember that the volume of a gas depends on the temperature and pressure. An increase in temperature causes a gas to expand significantly and this could cause an error in this type of experiment.

- To investigate the rate, the volume of gas is measured against time and a graph of gas volume against time is plotted.
- The gas in a syringe will always contain some air pushed into the syringe by the gas that is generated in the reaction. This will not affect the result since the total volume of gas and air collected in the syringe will be identical to the volume of gas produced.

Tip

Remember that if a solution has to be added to a substance in the flask to generate the gas, a source of error is that the gas could be lost through the top of the flask before the bung has been replaced. See p. 14.

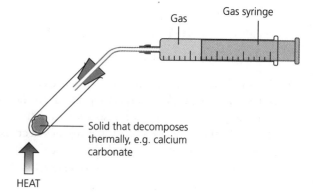

Figure 26 Heating a sample and collecting the gas. Note the reaction is over when the gas syringe stops moving

Knowledge check 30

In an experiment a gas could be collected in a $100\,cm^3$ measuring cylinder, $100\,cm^3$ gas syringe or $50.0\,cm^3$ burette. Which measuring instrument has higher resolution?

Measuring change in mass

A reaction in which a gas is produced may also be monitored by measuring the mass over a period of time. This method is not suitable for hydrogen, which has a very low formula mass and the mass lost would be small and difficult to measure. The apparatus is shown in Figure 27. A graph of mass against time is often plotted.

Tip

If a reaction is highly exothermic, it is difficult to accurately determine the rate because the higher temperature causes an increase in reaction rate, which causes the reaction to proceed faster.

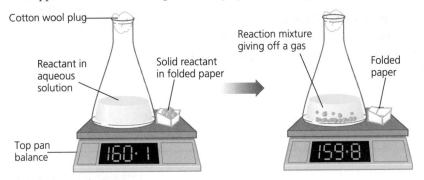

Figure 27 Following the course of a reaction by measuring the change of mass

Measuring change in a reactant or product by titration

Sometimes samples of the reacting mixture are taken at various times, during the course of a reaction and the reaction is **quenched**. This means the reaction is stopped. Methods of quenching include rapid cooling, adding a chemical to remove a reactant which is not being monitored or adding a large known volume of water to the sample. Each sample may then be titrated to find the concentration of the reactant or product and a graph of concentration against time drawn. See Figure 28.

Tip

The shape of the graph, and the calculation of half-life can be used to determine the order. The gradient of tangents at different concentrations can give the rate and a rate against concentration graph also used to find the order.

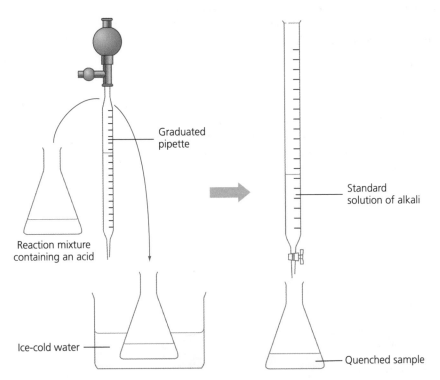

Figure 28 Following the course of a reaction during which there is a change in concentration of acids present by removing measured samples of the mixture at intervals, quenching, then determining the concentration of one reactant or product by titration

Measuring a coloured reactant or product

Measurement of a coloured reactant or product can be used for continuous rate monitoring. A **colorimeter** measures the colour intensity of a solution. A calibration curve should be set up first with known concentrations of the reactant or product, so that colorimeter readings are related to concentration.

In the continuous rate method, one experiment is carried out and the colorimeter readings are converted to concentration using the calibration curve. A graph of concentration against time is drawn and the shape of this curve can give the order with respect to the coloured reactant.

Worked example

Suggest a suitable method for measuring the rate of each of these reactions:

a $CH_3COOCH_3(l) + H_2O(l) \rightarrow CH_3COOH(aq) + CH_3OH(aq)$

b $C_4H_9Br(l) + H_2O(l) \rightarrow C_4H_9OH(l) + H^+(aq) + Br^-(aq)$

c $MgCO_3(s) + 2HCl(aq) \rightarrow MgCl_2(aq) + CO_2(g) + H_2O(l)$

Answers

a This is hydrolysis of an ester and there will be a change in concentration of ethanoic acid as it is produced in the reaction. Remove samples at intervals, quench the reaction by cooling and then titrate against alkali to determine the concentration of acid at different times.

b Use a pH meter to measure the changes in conductivity as the number of hydrogen ions increases.

c A gas, carbon dioxide, is produced. Carry out the reaction in a flask, with a loose plug of cotton wool in its neck, on a balance and record the loss in mass at regular intervals. Alternatively, measure the volume of carbon dioxide in a gas syringe at regular intervals against time.

Finding the order of reaction

It is possible to find the order for a particular reagent by the continuous monitoring method. This method works if the reaction involves only one reagent. Alternatively, if more reagents are present, the reaction can be set up so that the reaction rate depends solely on one reagent, and all the other reagents are present in such a *large excess* that their concentration is *nearly constant* and so they have little effect on the reaction rate. The order can be determined from the shape of the concentration–time graph (Figure 29).

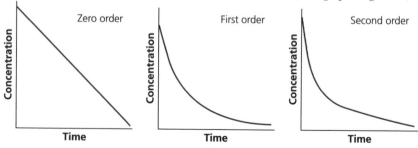

Figure 29 Concentration–time graphs

Tip

An appropriate coloured filter is used with a colorimeter. For example, if you are measuring the intensity of a blue solution, you should use a red filter as the solution is absorbing red light. The amount of red light absorbed relates directly to the concentration.

Knowledge check 31

Magnesium reacts with copper(II) sulfate. What method could be used to determine the effect of the concentration of copper(II) sulfate on the rate of reaction?

A zero-order reaction has a straight-line graph. For a first-order graph the graph of concentration against time is a curve. From this graph the **half-life** of the reaction can be worked out. If the *half-life is constant,* then the reaction is *first order*.

Figure 30 shows a concentration–time graph. From the graph you can see that:
- time taken for concentration to fall by half from 1.0 to 0.5 is 100 s
- time taken for concentration to fall by half from 0.5 to 0.25 is 100 s
- time taken for concentration to fall by half from 0.25 is 0.125 is 100 s

Hence the half-life is constant at 100 s proving the graph is for a first-order reaction.

The **half-life** of a reaction is the time taken for the concentration of one of the reactants to fall by half.

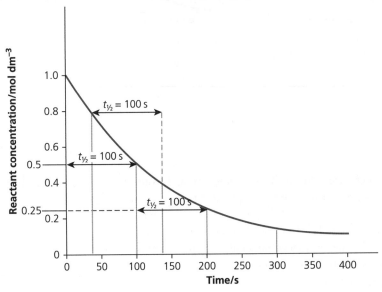

Figure 30 A constant half-life shows that the reaction is first order

The graph of concentration against time for a second-order reaction is shown in Figure 31. The half-life for a second-order reaction is not constant. Take the gradients of tangents taken at points to find the rate.

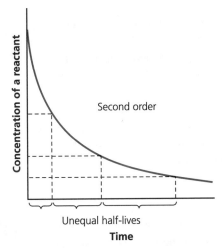

Figure 31 A graph of concentration against time for a second-order reaction

Tip

If the half-lives show that a reaction is not first order, it may be second order, but this needs to be checked by finding the rate at different points. Some reactions may have a fractional order.

Skills Guidance

Another way of determining order is to find the rate at different concentrations by finding the gradient of tangents at these points and plotting a rate against concentration graph. The shape gives the order (Figure 32).

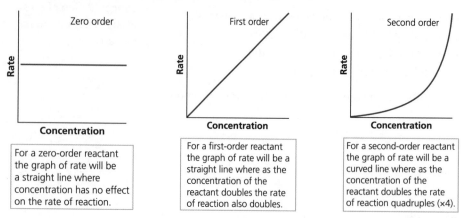

For a zero-order reactant the graph of rate will be a straight line where concentration has no effect on the rate of reaction.

For a first-order reactant the graph of rate will be a straight line where as the concentration of the reactant doubles the rate of reaction also doubles.

For a second-order reactant the graph of rate will be a curved line where as the concentration of the reactant doubles the rate of reaction quadruples (×4).

Figure 32 Rate against concentration graphs

Worked example

Calcium carbonate reacts with hydrochloric acid. In this reaction carbon dioxide is produced. If the calcium carbonate is present in excess, it is possible to investigate the rate of this reaction and determine the order with respect to the acid. The rate of reaction can be monitored by monitoring the loss in mass. A conical flask containing reactants with a cotton wool plug in the neck is placed on a balance and the mass is recorded every 10 seconds initially and then every 20 seconds.

a What is the purpose of the cotton wool plug?

b What is observed in the flask during the reaction?

c Why is the time interval of recording changed?

d When do you stop taking readings in this experiment?

e Suggest why this reaction does not go to completion in the time available for the experiment.

f Write a balanced symbol equation, with state symbols for the reaction which occurs in this experiment.

g Another method of monitoring the rate of this reaction is by gas collection. The gas can be collected underwater in an inverted measuring cylinder. State a source of error in this experiment.

h Collecting the gas in a $100\,cm^3$ gas syringe is a more effective method of gas collection. Show by calculation that a $100\,cm^3$ gas syringe is suitable to collect the gas produced by the reaction of $10\,g$ (excess) of calcium carbonate with $15.0\,cm^3$ of $0.5\,mol\,dm^{-3}$ HCl(aq) at room temperature and pressure.

Answers

a To prevent loss of acid spray.

b Bubbles, the calcium carbonate disappears and a colourless solution is formed.

c The reaction rate is fastest at the beginning and so it is best to record readings more often.

Knowledge check 32

What is meant by the term 'half-life'?

48   OCR Chemistry

d When two readings are constant, showing that no more carbon dioxide is being lost and the reaction is over.

e The HCl(aq) is used up in the reaction and so the concentration of HCl(aq) decreases so much that the reaction becomes so slow that it seems to stop, even though it may not have gone to completion.

f $CaCO_3(s) + 2HCl(aq) \rightarrow CaCl_2(aq) + H_2O(l) + CO_2(g)$

g Carbon dioxide may dissolve in the water.

h moles of HCl $= \dfrac{\text{volume (cm}^3) \times \text{concentration}}{1000} = \dfrac{15.0 \times 0.5}{1000} = 0.0075$

ratio 2 moles HCl : 1 moles CO_2

moles $CO_2 = 0.00375 = \dfrac{\text{vol (cm}^3)}{24\,000}$

volume $= 90\,cm^3$

PAG10 Rate of reaction: initial rates method

In this method, the rate is determined immediately after the start of the reaction. At this point all of the concentrations are known. A series of experiments is carried out in which the initial concentration of one of the reactants is varied systematically while the concentrations of the other reactants are kept constant. The time, t, is measured to a fixed point in the reaction. In the method of initial rates, it is necessary to determine the initial rate for each experiment. Often results are plotted on a concentration–time graph. Initial rates can be determined from the concentration–time graph. A tangent is drawn at $t = 0$ and the rate is the gradient of this tangent.

Tangents to a curve

The word tangent means 'touching' in Latin. The tangent is a straight line which just touches the curve at a given point and does not cross the curve.

To draw a tangent at a point (x, y), see Figure 33.

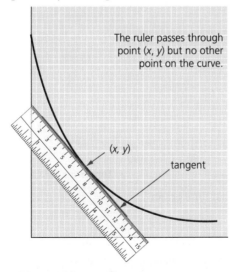

The ruler passes through point (x, y) but no other point on the curve.

(x, y)

tangent

Figure 33 Drawing a tangent to a curve

Skills Guidance

- Place your ruler through the point (x, y).
- Make sure your ruler goes through the point and does not touch the curve at any other point.
- Draw a ruled pencil line passing through point (x, y).

To calculate the *gradient* of a curve at a particular point, it is necessary to draw a tangent to the curve at the point and calculate the gradient of the tangent.

Knowledge check 34

What are the units of the gradient for a graph of rate against concentration?

Worked example

A student recorded the total volume of gas collected in a reaction at 20-second intervals and plotted the graph shown in Figure 34. Use the graph to calculate the rate of reaction at 60 s. State the units.

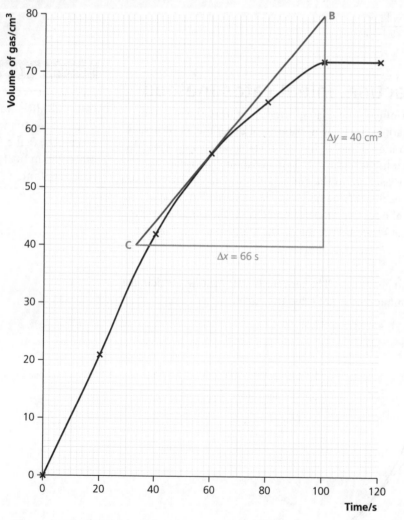

Figure 34 Volume of gas against time graph

Answer

First draw the tangent at time 60 s — this is the red line on the graph. The rate at this point is given by the gradient of this tangent. To find the gradient, choose two points, B and C, far apart on the line and form a triangle as shown in green on Figure 33.

$$\text{gradient } (m) = \frac{\text{change in } y\text{-axis}}{\text{change in } x\text{-axis}} = \frac{\Delta y}{\Delta x} = \frac{40}{66} = 0.61 \text{ (2 s.f.)}$$

$$\text{units} = \text{cm}^3\text{s}^{-1}$$

To find the slope of the curve at any other point, draw a tangent line at that point and then determine the slope of that tangent line.

■ If the concentrations of the reagents are known at the start of the reaction, a measurement of the initial gradient ($t = 0$) gives a numerical value of the rate at these concentrations.

■ If the experiment is repeated with the concentration of one of the reactants doubled, while the other concentrations are kept the same, a new initial rate can be established. Therefore, it is possible to see the effect of this reactant on the overall rate. If it is found that the gradient has doubled, then the reaction is first order with respect to that reactant. If, however, the rate increased four fold, it would be second order with respect to the reagent. Changing the concentrations of each reactant in turn allows the order of reaction with respect to each reagent to be determined.

Clock reactions

To carry out several full experiments using the methods described on pp. 44–46 to provide data for the initial rate of a reaction is time consuming. Often, instead, an approximation to obtain initial rate is used.

■ The time taken to reach a specific point in the reaction soon after the reaction has started, can be recorded and the reaction repeated with different concentrations to determine how the time taken to reach this point changes. The rate to this point is taken to be proportional to 1/time (1/t) for each reaction.

Clock reactions are ideal for measuring the time for a certain amount of product to be formed. A clock reaction measures the time from the start of the reaction until there is a visual change (preferably sudden) such as:

■ appearance of a precipitate
■ disappearance of a solid
■ change in colour

The reaction of hydrogen peroxide with iodide ions in acid solution can be set up as a clock reaction:

$$H_2O_2(aq) + 2H^+(aq) + 2I^-(aq) \rightarrow I_2(aq) + 2H_2O(l)$$

A small, known amount of sodium thiosulfate ions is added to the reaction mixture, which also contains starch indicator. At first the thiosulfate ions react with any iodine, I_2, as soon as it is formed, turning it back to iodide ions, so there is no colour change. At the instant when all the thiosulfate ions have been used up, free iodine is produced and this immediately gives a deep blue-black colour with the starch. The reaction time t for the initial part of the reaction to take place is measured and $1/t$ calculated as measure of the initial rate of a reaction. The experiment can be repeated with different initial concentrations of the reactants and the time taken to reach the blue-black colour recorded.

Tip

It is assumed that the reaction proceeds at a constant rate to the point that it is measured. This is not strictly true because, as soon as the reaction starts, the rate begins to slow. However, the approximation is good enough to determine an integer order.

Knowledge check 35

In a reaction doubling the concentration of hydrogen doubles the rate of reaction when the concentration of other reactants is constant. Tripling the concentration of NO gas increases the rate by a factor of nine, when other reactants have constant concentration. Write the rate equation for the reaction.

Worked example

The graph shown in Figure 35 is for an iodine clock reaction. The blue-black colour appeared at 60 seconds. Explain, using the graph, why the estimate of the initial rate of a reaction determined by this clock reaction is close to, but not equal to, the true initial rate.

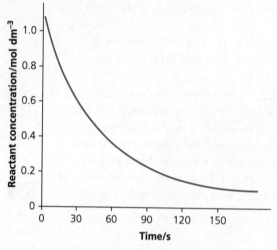

Figure 35 A concentration–time graph

Answer

The graph gradient decreases, showing that the reaction proceeds quickly at the start and slows down towards the finish, so the rate when the clock reaction ends is slower than the initial rate. A tangent to the curve is drawn at $t = 0$ and at $t = 60$ seconds (blue lines), and the gradient calculated.

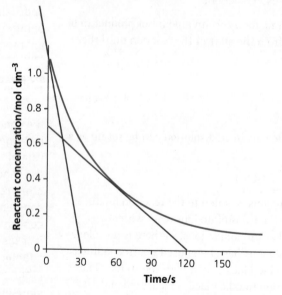

From the graph:

- initial rate, at time $t = 0$ seconds, is $\dfrac{1.07}{30} = 0.036 \, \text{mol} \, \text{dm}^{-3} \, \text{s}^{-1}$
- rate after 60 seconds is $\dfrac{0.7}{120} = 0.006 \, \text{mol} \, \text{dm}^{-3} \, \text{s}^{-1}$

These values show that the rate slows considerably as the reaction progresses and at time 60 seconds, when the colour change of the clock reaction occurs, the rate is much slower as the reactants have started to be used up. The rate is at its initial maximum before the reactants have started to be used up. Hence the clock reaction is an approximation and works best if the reaction has not progressed very far.

Worked example

The iodate(v) ion is an oxidising agent and reacts with sulfate(ıv) ions in acidic solution to produce iodine in solution, according to the following equations:

> Stage 1: $IO_3^- + 3SO_3^{2-} \rightarrow I^- + 3SO_4^{2-}$
>
> Stage 2: $IO_3^- + 5I^- + 6H^+ \rightarrow 3H_2O + 3I_2$
>
> Overall: $2IO_3^- + 5SO_3^{2-} + 2H^+ \rightarrow I_2 + 5SO_4^{2-} + H_2O$

Iodine is only liberated when acid is added. It is possible to determine the effect of changing the concentration of acid on the initial rate of this reaction by timing how long it takes for iodine to be produced.

a What should be added to the reaction mixture to indicate when iodine has been produced?

b The reactants were placed in a beaker and the timer started, and the contents stirred with a glass rod. Why were the contents stirred?

c When is the timer stopped?

d The concentration of acid was changed in this experiment. Name two controlled variables in the experiment.

e The experiment was designed to determine the order of reaction with respect to hydrogen ions in the reaction. What changes would you make to the experiment so that the order of reaction with respect to iodate(v) ions could be determined?

f The reaction is first order with respect to hydrogen ions. In an experiment to determine the order of this reaction, a value of 0.963 was obtained. Calculate the percentage error in this result.

g The experimental error resulting from the use of the apparatus was determined to be 2.1%. Explain what this means in relation to the practical technique used.

Answers

a Starch

b To ensure the reactants all mix and react

c When the blue-black colour of starch appears

d The volume of all solutions/the concentrations of the other solutions/temperature

e Keep the concentration of acid constant and vary the concentration of iodate(v)

→

f Error is $1 - 0.963 = 0.037$

$$\% \text{ error} = \frac{0.037}{1} \times 100 = 3.7\%$$

g The result error is greater than the apparatus error, so the fault is due to the person carrying out the experiment.

In a clock experiment reacting sulfuric acid, potassium iodate(v), and sodium sulfate(IV) solutions, in the presence of starch, all of the concentrations were kept constant except the concentration of acid, hence the rate equation for the reaction:

$$\text{rate} = k[IO_3^-]^x[SO_4^{2-}]^y[H^+]^z$$

simplifies to:

$$\text{rate} = k[H^+]^z$$

where z is the order of reaction with respect to hydrogen ion concentration. This expression can be written:

$$\log(\text{rate}) = z\log[H^+] + \text{constant}$$

The event being measured is fixed — that is, the first appearance of the blue–black colour — so it is possible for the rate to be expressed as:

$$\text{rate} = \frac{1}{t}$$

The total volume of solution in each experiment is constant and so the $[H^+]$ can be represented by the volume of sulfuric acid. Therefore the rate expression becomes:

$$\log\left(\frac{1}{t}\right) = z\log(\text{volume of sulfuric acid}) + \text{constant}$$

The results are given in Table 17. $\frac{1}{t}$ has been calculated as a measure of the rate, and $\log\left(\frac{1}{t}\right)$ and $\log(\text{volume of sulfuric acid})$ are also calculated.

Tip

You do not need to be able to deduce this expression, but you should be able to calculate logs, as shown in Table 17.

Table 17

Volume acid/cm³	25	35	45	55	70	85
t/s	42	28	10	15	11	9
$1/t$/s⁻¹	0.0238	0.0357	0.500	0.677	0.0909	0.111
$\log(1/t)$	−1.62	−1.40	−1.30	−1.18	−1.04	−0.96
$\log(\text{vol})$	1.40	1.54	1.65	1.74	1.85	1.93

A graph of of $\log\left(\frac{1}{t}\right)$ (y-axis) against $\log V$ (volume) (x-axis) is shown in Figure 36. It is a straight line of gradient z and from this the order with respect to hydrogen ion concentration can be determined. The graph is of the type $y = mx + c$ and therefore the gradient m will give a value for the order, z.

The gradient is $0.60/0.50 = 1.2$. Hence the reaction is first order with respect to hydrogen ion concentration.

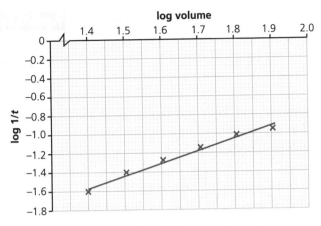

Figure 36 Graph of log (1/t) against log volume

PAG11 pH measurement

pH of solutions can be measured with a pH meter, pH probe, data logger or narrow-range pH paper. The pH change in a solution during an acid–base titration can be measured using a pH probe. Periodic measurement of the pH of the reaction solution allows plotting of a graph of pH against volume of solution added. Such a graph is called a pH titration curve and each has a characteristic shape for the various combinations of strong and weak acids and bases (see Figure 37). (**Note for spec B you do not need to recall the shape of these curves.**)

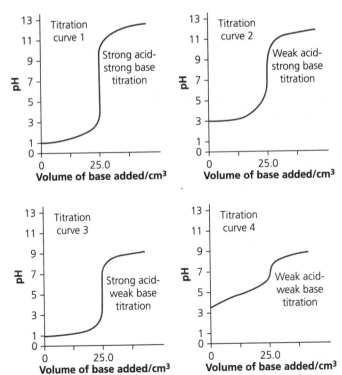

Figure 37 Titration curves

Skills Guidance

When using a pH meter or probe, it is necessary to calibrate it before use, so that accurate pH values can be obtained.

Worked example

a Methanoic acid is a weak acid. In an experiment, a calibrated pH meter was used to measure the pH of methanoic acid solution. At 20°C the pH of a $0.100\,mol\,dm^{-3}$ solution was 2.37.

  i Explain why a pH meter should be calibrated before use.

  ii Explain how a pH meter is calibrated.

  iii Write an equation for the dissociation of methanoic acid and explain what is meant by weak acid.

  iv Write an expression for the equilibrium constant, K_a, for the dissociation of methanoic acid in aqueous solution and calculate the value of K_a for this dissociation at 20°C. Give your answer to the appropriate precision.

b A student used aqueous sodium hydroxide to determine the titration curve for the reaction of methanoic acid and sodium hydroxide solution. $25.0\,cm^3$ of $1.50 \times 10^{-2}\,mol\,dm^{-3}$ methanoic acid was placed in a beaker at 25°C. The sodium hydroxide was added in $2.0\,cm^3$ portions from a burette and the pH of the reaction mixture was measured using a pH meter.

  i Describe how $25.0\,cm^3$ of methanoic acid was accurately measured and placed in the conical flask.

  ii Describe how the burette was prepared.

  iii Write a balanced symbol equation for the reaction between HCOOH and NaOH.

  iv Why was the reaction mixture swirled after the addition of each $2.0\,cm^3$ portion of sodium hydroxide?

  v The pH curve for this titration is shown in Figure 38. Calculate the value of the concentration in $mol\,dm^{-3}$ of the aqueous sodium hydroxide.

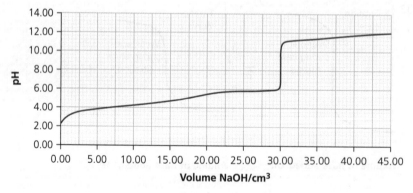

Figure 38 A titration curve

  vi The pH ranges in which the colour changes for three acid–base indicators are shown in Table 18. Explain which one of the three indicators is suitable for this titration (**not for spec B**).

Table 18

Indicator	pH range
Metacresol purple	7.40–9.00
2,4,6-trinitrotoluene	11.50–13.00
Ethyl orange	2.4–4.8

Answers

a i After storage, a pH meter does not give accurate readings because the glass electrode in the pH meter does not give a reproducible emf over longer periods of time.

ii Calibration should be performed with at least two standard buffer solutions that span the range of pH values to be measured. The pH probe is rinsed thoroughly with deionised water, and shaken to remove excess water. It is then placed in a standard buffer solution of one pH, ensuring the bulb is fully immersed and allowed to sit until the pH stabilises. Adjust the reading to the pH of the buffer. Rinse the probe with deionised water and repeat with a different pH buffer.

iii $HCOOH(aq) \rightleftharpoons HCOO^-(aq) + H^+(aq)$

The reversible arrow shows that the dissociation is incomplete since HCOOH is a weak acid.

iv $K_a = \dfrac{[COO^-][H^+]}{[HCOOH]}$

$pH = -\log[H^+]$

$2.37 = -\log[H^+]$

$[H^+] = 4.27 \times 10^{-3}$

$K_a = \dfrac{(4.27 \times 10^{-3})^2}{0.100}$

$K_a = 1.82 \times 10^{-4}\,mol\,dm^{-3}$

The answer is given to three significant figures, as that is the accuracy of the data given.

b i A $25.0\,cm^3$ pipette was rinsed with methanoic acid and then filled using a safety pipette filler, until the bottom of the meniscus was on the $25.0\,cm^3$ mark at eye level. It was transferred to the beaker and allowed to run out. The pipette was touched on the surface of the liquid to expel the last drops.

ii The burette was rinsed with sodium hydroxide solution, and then filled to above the zero mark. The solution was run down until the bottom of the meniscus was on the zero mark at eye level. This ensures that the tap is filled. The burette was checked to ensure there were no bubbles.

iii $NaOH + HCOOH \rightleftharpoons HCOONa + H_2O$

iv To ensure that the solution is homogeneous and the pH is uniform throughout.

v moles of methanoic acid $= \dfrac{25.0 \times 1.5 \times 10^{-2}}{1000} = 3.75 \times 10^{-4}$

ratio $1:1$

moles NaOH $= 3.75 \times 10^{-4} = \dfrac{30.0 \times concentration}{1000}$

conc $= 0.0125\,mol\,dm^{-3}$

vi The pH change at the end point (vertical portion) is between 6 and 11. Metacresol purple changes colour within this range and is suitable to use.

Tip

An alternative technique for calibrating is to record the pH reading for each buffer and plot a graph of the pH of the recorded pH (x-axis) against the pH of the buffer solution. This calibration curve can be used to convert pH readings in an experiment into more accurate values.

Questions & Answers

Both OCR specifications for the A-level examination consist of three papers. All papers will examine practical skills in chemistry.

In specification B, although all papers may include questions in a practical context, paper 3 will have a particular emphasis on practical skills, including data analysis. All the following questions are suitable for specifications A and B.

This student guide covers only practical work, mainly through consideration of the 11 practical activity groups. Therefore, the questions in this section only reflect the type of question that will be set to *test your understanding of experimental methods* and are *not* representative of the papers as a whole.

In the AS examination there are just two papers and questions based on practical work may be set in both papers. **If you are studying for AS then concentrate on questions 1–11 in this section**.

A data sheet is provided with each examination. Copies may be downloaded from the OCR website.

Questions on all papers are a mixture of multiple choice, short answer and longer answer structured questions.

You should pay particular attention to diagrams, drawing graphs and making calculations. Many students lose marks by failing to label diagrams properly, not giving essential data on graphs and, in calculations, by not showing all the working or by omitting units.

In this section, comments preceded by 🄔 indicate what to watch out for when answering the question. Comments on student answers are preceded by the icon 🄔.

Ticks (✓) are included in the answers to show where a mark has been awarded.

Question 1

Hydrated nickel(II) nitrate is not used in school laboratories due to its toxicity. The mass of water in $Ni(NO_3)_2.xH_2O$ (hydrated nickel(II) nitrate) was determined in an analytical laboratory, by gently heating a known mass of the solid in a fume cupboard to drive off all the water of crystallisation and reweighing.

(a) State how to ensure, using weighings, that all the water of crystallisation was removed in this experiment.

[1]

> **Student answer**
>
> (a) Weigh, heat, reweigh and repeat until the mass is constant. ✓

🄔 The substance should be weighed in an evaporating basin and heated and weighed repeatedly (heated to constant mass) — when the mass does not change all of the water of crystallisation has been removed.

(b) In an experiment 2.91 g of hydrated nickel(II) nitrate, $Ni(NO_3)_2.xH_2O$, produced 1.83 g of the anhydrous salt.

   (i) Write the equation for the reaction that occurs in the experiment. [1]

   (ii) Determine the value of x in $Ni(NO_3)_2.xH_2O$. [3]

ⓔ In this calculation you need to first determine the mass of water that has been removed, and then the mass of anhydrous nickel(II) nitrate left. Then work out the moles of each and put them in ratio to find x.

   (iii) Suggest one reason why the value of x determined by this experiment may be less than the actual value. [1]

   (iv) Hydrated nickel(II) nitrate is a green crystalline solid. State one other observation which may be made during the heating of the hydrated nickel(II) nitrate. [1]

   (v) Suggest what effect very strong heating may have on the salt. [1]

(b) (i) $Ni(NO_3)_2.xH_2O \rightarrow Ni(NO_3)_2 + xH_2O$ ✓

   (ii) mass of water = 2.91 − 1.83 = 1.08 g

$$\text{moles of water} = \frac{1.08}{18.0} = 0.0600 ✓$$

$$\text{moles } Ni(NO_3)_2 = \frac{1.83}{182.7} = 0.0100 ✓$$

ratio $Ni(NO_3)_2 : H_2O$

0.0100 : 0.0600

1 : 6, so $x = 6$ ✓

   (iii) All of the water may not have been removed ✓.

   (iv) Steam/beads of colourless liquid/crystals change to powder ✓.

   (v) It may cause it to decompose ✓.

ⓔ The salt is a nitrate and if nitrates are heated they may thermally decompose to give the metal oxide, nitrogen dioxide and oxygen.

Question 2

(a) In an experiment to compare the percentage of magnesium carbonate in two different rocks, powdered rock samples were heated and the carbon dioxide produced collected.
Describe how this experiment is carried out. Include a labelled diagram. Assume that the rocks contain no other carbonates. [4]

ⓔ Because this experiment is a comparison, it should be a fair test. The powdered rock could be heated in a test tube and the gas collected in a gas syringe or under water in a measuring cylinder.

> **(a)** Comment on fair test — equal masses of rock/similar heating conditions and length of time heating. ✓
>
> Labelled diagram to show rock being heated in a vessel with one exit ✓ via delivery tube ✓ either a gas syringe or measuring cylinder under water ✓.

(b) Without carrying out calculations, how could your results show which rock had the highest percentage magnesium carbonate present? [1]

> **(b)** More $MgCO_3$ will give off more CO_2. ✓

Question 3

Copper(II) chloride crystals can prepared by neutralising dilute hydrochloric acid with an excess of solid copper(II) oxide or carbonate.

(a) (i) Write the equation, with state symbols, for the reaction of copper(II) carbonate with hydrochloric acid. [2]

(ii) Describe the steps in the method to prepare a sample of pure dry copper(II) chloride $CuCl_2.2H_2O$ from insoluble copper(II) carbonate and dilute hydrochloric acid (without details of the apparatus used). [4]

> **(a) (i)** $CuCO_3(s) + 2HCl(aq) \rightarrow CuCl_2(aq) + H_2O(l) + CO_2(g)$ ✓✓
>
> **(ii)** Add $CuCO_3$ to acid until excess is left/no more bubbles and acid is used up. ✓
>
> Filter (to remove the excess carbonate). ✓
>
> Evaporate to half volume. ✓
>
> Filter the crystals and dry between filter paper. ✓

ⓔ To prepare a salt, excess solid copper(II) carbonate must be added to the acid, to ensure that it is all used up. The excess copper(II) carbonate must be removed by filtration and then the solution evaporated to half volume only, to ensure water of crystallisation is not removed.

(b) The percentage yield was much less than 100%. Some explanations for this are:

   A   The crystals are damp at the end.

   B   Some copper carbonate remains unreacted at the end.

   C   The copper(II) chloride has lost some of its $2H_2O$ water of crystallisation.

For each explanation, state whether it could be correct and explain your answer. [3]

> **(b) A**   No — this would have resulted in a higher mass/yield. ✓
>
>   **B**   No — copper(II) carbonate is added in excess, and it is the moles of hydrochloric acid that are the limiting factor. ✓
>
>   **C**   Yes — loss of water would reduce mass. ✓

ⓔ Water present in the crystals would result in a higher mass of product. Losing some water of crystallisation would result in a lower mass of product. The copper carbonate is added in excess and there will be some unreacted at the end; it will not affect the mass of product.

(c) What is observed when a few drops of silver nitrate are added to a solution of copper chloride? [1]

(c) White ppt ✓

ⓔ Adding silver nitrate is the test for chloride ions. A white precipitate is observed in the blue solution. An alternative question for A-level would be to ask what is observed when sodium hydroxide solution is added in excess — a blue precipitate.

Question 4

A solution of a metal hydroxide XOH was made up by dissolving 3.92 g of solid in 250 cm^3 of water. 25.0 cm^3 of XOH solution was placed in a conical flask with a few drops of bromothymol blue indicator. The indicator is yellow in acid and blue in alkali. The conical flask was placed on a white tile and titrated with a standard solution of hydrochloric acid of concentration 0.500 mol dm^{-3}, in order to find the concentration of XOH.

The balanced symbol equation for the reaction is:

$$XOH + HCl \rightarrow XCl + H_2O$$

(a) Describe how you would safely and accurately measure out and place 25.0 cm^3 of the metal hydroxide solution in the conical flask, naming all the apparatus you would use. [3]

(a) Rinse a pipette with the XOH solution. ✓

Using a pipette ✓ and pipette filler ✓ transfer 25.0 cm^3 of this solution into a conical flask.

ⓔ To measure out 25.0 cm^3 accurately a pipette is used, and a safety pipette filler for safety. For accuracy, rinse the pipette with the XOH.

(b) Why is a white tile used in this practical technique? [1]

(b) To aid detection of the colour change ✓

(c) State the colour change of the bromothymol blue indicator at the end point. [1]

(c) blue to yellow ✓

ⓔ XOH is an alkali and is in the conical flask, so the colour change will be from blue to yellow.

(d) Describe how the end point was accurately determined. [1]

> **(d)** Add acid drop-wise, ✓ with swirling until the colour changes.

> ⓔ To determine the end point accurately, after a trial titration has been carried out, run the solution in until near the trial value, and then add in drops, with swirling.

(e) Why is the conical flask swirled during a titration? [1]

> **(e)** To mix the solution and ensure the solutions have reacted. ✓

(f) The burette was not filled correctly and the gap between the tap and the tip of the burette still contained air. Suggest the effect this would have on the titre. [1]

> **(f)** The titre would be greater. ✓

> ⓔ When the burette is opened, the tap is filled before any liquid is delivered into the flask. Hence the measured volume would be greater.

The results obtained in this experiment are shown below.

	Trial	1	2	3
Final burette reading/cm^3	14.40	28.25	42.35	14.30
Initial burette reading/cm^3	0.00	14.30	28.25	0.00
Volume of HCl used/cm^3	14.40	13.95	14.10	14.30

(g) Calculate the concentration of the XOH in mol dm^{-3}. Give your answer to three significant figures. [3]

> ⓔ First calculate the average titre using concordant values — 13.95 and 14.10. Then calculate the amount in moles of acid and by ratio the moles of XOH.

> **(g)** average titre = $\dfrac{13.95 + 14.10}{2}$ = 14.0 ✓
>
> moles HCl = $\dfrac{14.0 \times 0.5}{1000}$ = 0.007
>
> ratio 1 : 1 moles XOH = 0.007
>
> $\dfrac{0.007}{0.025}$ = 0.28 mol dm^{-3} ✓✓

(h) Calculate the relative formula mass (M_r) of XOH and identify the element X. [3]

> **(h)** From the initial weighings 3.92 g in 250 cm^3
>
> 3.92 = 15.68 g dm^{-3}
>
> $\dfrac{15.68\,\text{g dm}^{-3}}{0.28\,\text{mol dm}^{-3}}$ = 56, XOH = 56 ✓✓ X = 56 − 17 = 39. Element X is K ✓

(i) A student obtained a result for X and her teacher told her it was lower than
expected. Suggest a reason for this.

[1]

> **(i)** The solid XOH may absorb water from the air/The XOH may react with
> carbon dioxide in the air. ✓

Question 5

(a) Describe how a value for the enthalpy change of combustion of ethanol can
be determined experimentally.

[6]

ⓔ Ethanol is a liquid and the method will involve placing ethanol in a spirit burner
and heating a known mass of water. You need to give clear and logical steps in
your answer and ensure you describe how the results are used to determine the
enthalpy change.

> **(a)** Find temperature change of water ✓
>
> Use a known volume of water in a calorimeter. ✓
>
> Burn measured mass of fuel in a spirit burner. ✓
>
> Measure temperature of water using thermometer. ✓
>
> Calculate number of moles of fuel used using mole $= \dfrac{mass}{M_r}$. ✓
>
> Use $q = mc\Delta T$ and scale up for 1 mole of fuel. ✓

(b) When the enthalpy of combustion of pentan-1-ol is determined, its value is
much larger. Suggest a reason for this.

[1]

ⓔ Each successive member of the homologous series adds an extra $-CH_2$ group.
Hence pentan-1-ol has more bonds to be broken and made during combustion,
and so there is a greater value.

> **(b)** More bonds to be broken and made ✓

(c) The value for the enthalpy change of combustion determined experimentally
is frequently much less exothermic than the value found in data books. One
reason for this is heat loss. Suggest two other reasons.

[2]

> **(c)** Incomplete combustion ✓
>
> Evaporation of fuel from wick of the burner ✓
>
> Not carried out under standard conditions. ✓ (any 2)

Question 6

When an ionic solid dissolves in water, there may be a temperature change. A student placed 100 cm³ of water in a polystyrene beaker and then recorded the temperature. He powdered some hydrated copper(II) sulfate, $CuSO_4.5H_2O$, and then dissolved approximately 2.10 g of it in the water and recorded the final temperature.

Mass of hydrated copper(II) sulfate added = 2.07 g

Initial temperature = 18.0°C

Final temperature = 17.8°C

Specific heat capacity of water is $4.18 \, J \, °C^{-1} \, g^{-1}$

(a) Why and how is the hydrated copper(II) sulfate powdered? [2]

> (a) Dissolves faster ✓
>
>   Using a mortar and pestle ✓

ⓔ A mortar and pestle is used to grind up solids. A powder has a greater surface area and reacts or dissolves faster.

(b) Describe, stating the equipment used, how the student could accurately determine the mass of hydrated copper(II) sulfate added. [3]

> (b) Use a top-pan balance reading to two decimal places. ✓
>
>   Weigh solid in weighing boat, add solid to water and reweigh weighing boat. ✓
>
>   Subtract values. ✓

ⓔ The recorded value is to two decimal places, so a top-pan balance which reads to two decimal places must be used. When using a balance, it is best to measure the solid in a weighing boat, add to the reaction, and then reweigh the weighing boat and subtract the values.

(c) Calculate the enthalpy change on dissolving the hydrated copper(II) sulfate in the water. [1]

> (c) Energy absorbed by water = $q = mc\Delta T = 100 \times 4.18 \times 0.20 = 84.0 \, J$ ✓

ⓔ When working out the enthalpy change it is always the mass of water which is used — 100 cm³ of water has a mass of 100 g since the density of water is $1 \, g \, cm^{-3}$. The temperature *change* can be used in °C; there is no need to convert to kelvin. The units of enthalpy change are joules.

(d) Calculate the enthalpy change, in $kJ \, mol^{-1}$, on dissolving 1 mole of hydrated copper(II) sulfate in water. [3]

ⓔ When calculating the M_r of hydrated copper(II) sulfate, remember to include the five moles of water of crystallisation.

(d) $M_r = 63.5 + 32.1 + (4 \times 16.0) + (5 \times 18) = 249.6$ ✓

Moles of hydrated copper(ii) sulfate used $= \dfrac{2.07}{249.6} = 0.00829$ ✓

0.00829 mol produces 84.0 J

1 mol produces $\dfrac{84}{0.00829} = 10132.69\,J\,mol^{-1} = 10.133\,kJ\,mol^{-1}$ ✓

$\Delta H = -10.1\,kJ\,mol^{-1}$

(e) The actual value is $-11.7\,kJ\,mol^{-1}$. State one source of error in the student's experiment and suggest how it could be reduced. [2]

(e) There are energy transfers to and from the surroundings. ✓

Use a lid/Use more insulation. ✓

Question 7

Some hydrochloric acid was added to some sodium carbonate solution in one test tube and silver nitrate in another test tube. Which letter gives the correct observations? [1]

	A	B	C	D
Sodium carbonate solution	No change	No change	Effervescence	Effervescence
Silver nitrate solution	Precipitate	No change	No change	Precipitate

The answer is D. ✓

e When an acid is added to a carbonate, effervescence results. The anion in hydrochloric acid is chloride and it will form a precipitate with silver nitrate solution.

Question 8

Which of the following is a correct statement about the test for bromide ions in solution? [1]

A Add $AgNO_3$(aq), white precipitate formed, soluble in dilute ammonia solution

B Add $AgNO_3$(aq), cream precipitate formed, soluble in dilute ammonia solution

C Add $AgNO_3$(aq), white precipitate formed, soluble in concentrated ammonia solution

D Add $AgNO_3$(aq), cream precipitate formed, soluble in concentrated ammonia solution

The answer is D. ✓

e Bromide ions form a cream precipitate with silver nitrate solution. The cream precipitate is not soluble in dilute ammonia but is soluble in concentrated ammonia solution.

Question 9

(a) A solid compound was dissolved in deionised water. Dilute nitric acid was added, followed by silver nitrate solution. A white precipitate was observed, which dissolved when ammonia solution was added.

(i) Name the ion that has been identified during this test. [1]

(ii) Write the simplest ionic equation for the formation of the white precipitate. [1]

(iii) Explain why dilute nitric acid is added to the solution before silver nitrate solution. [2]

> (i) Chloride ✓
>
> (ii) $Ag^+ + Cl^- \rightarrow AgCl$ ✓
>
> (iii) Nitric acid reacts with/removes carbonate/hydroxide ions. ✓
> Carbonate ions would give a white precipitate/false test for chloride ions. ✓

📝 Silver nitrate solution is the key to this question as it is used to test for halide (chloride, bromide and iodide) ions. Chloride ion gives a white precipitate. Be careful that you do not write down chlorine ion — this is incorrect. The dilute nitric acid is added to remove any carbonate ions, which would give a white precipitate of silver carbonate, Ag_2CO_3. This would be a false positive test for chloride ions. Learn the observations carefully in terms of the colour of the precipitates and whether they redissolve in ammonia solution.

(b) It is possible to distinguish between $BaCl_2$(aq) and $MgCl_2$(aq) by adding an aqueous reagent to a sample. Give a suitable aqueous reagent that could be added separately to each compound. Describe what you would observe in each case. [3]

📝 These are both group 2 compounds, You need to think about identification tests and about the solubility of group 2 compounds.

> (b) Add some aqueous sodium sulfate or sulfuric acid (any named soluble sulfate). ✓
>
> A white precipitate forms for barium chloride solution. ✓
>
> Whereas the solution remains colourless/No reaction for magnesium chloride solution ✓ or
>
> Add sodium or potassium hydroxide solution. ✓
>
> The solution remains colourless/No reaction for barium chloride solution. ✓
>
> A white precipitate occurs for magnesium chloride solution. ✓

Ⓔ Remember that magnesium hydroxide is insoluble but magnesium sulfate is soluble, and barium hydroxide is soluble but barium sulfate is insoluble. Adding any soluble sulfate, for example sodium sulfate solution or sulfuric acid, to the compounds will form magnesium sulfate, which is soluble and produces a colourless solution, and barium sulfate, which is insoluble and produces a white precipitate. Note that you must give the full name of the reagent — it is not sufficient to write 'add a soluble sulfate'. Alternatively, sodium or potassium hydroxide solution could be added. Magnesium hydroxide is formed as an insoluble white precipitate; barium hydroxide is soluble and a colourless solution remains.

Question 10

The following method can be used to prepare 1-bromobutane.

Place $30\,cm^3$ of water, 40 g of powdered sodium bromide and 21.8 g of butan-1-ol in a $250\,cm^3$ round-bottomed flask. Add $25\,cm^3$ of concentrated sulfuric acid dropwise, cooling the flask in an ice bath. When the addition is complete, reflux for 45 minutes and then distil off the crude 1-bromobutane. Wash the distillate with concentrated sulfuric acid and then with sodium carbonate solution. Remove the 1-bromobutane layer and add a spatula of anhydrous sodium sulfate, swirl and filter. Distil and collect the 1-bromobutane at 99–102°C.

(a) Write the equation for the formation of 1-bromobutane from butan-1-ol and hydrogen bromide.

[1]

Ⓔ This particular reaction is not on your specification but simply use the information in the question. You know the formula of butan-1ol — $CH_3CH_2CH_2CH_2OH$ — and of 1-bromobutane — $CH_3CH_2CH_2CH_2Br$. Put these into an equation.

(a) $CH_3CH_2CH_2CH_2OH + HBr \rightarrow CH_3CH_2CH_2CH_2Br + H_2O$ ✓

(b) What is reflux?

[1]

(b) Repeated boiling and condensing of a reaction mixture to ensure that reaction takes place without the contents of the flask boiling dry. ✓

(c) State three changes made to the apparatus set-up to change from reflux to distillation.

[3]

Ⓔ Reflux means that the condenser is in the vertical position. For distillation the condenser is sideways, and a still head connector and receiver need to be used. You should be able to draw this apparatus and describe it.

(c) Add a still head/Add a thermometer/Condenser moved from upright to sideways position/Receiver on end of condenser Any 3 ✓ ✓ ✓

Questions & Answers

(d) Suggest why sodium carbonate solution is added to wash the distillate, rather than sodium hydroxide solution. [2]

> **(d)** Sodium carbonate solution is a weaker alkali than sodium hydroxide solution. ✓
>
> Sodium hydroxide solution would hydrolyse the halogenoalkane. ✓

ⓔ First, think of the difference between sodium carbonate solution and sodium hydroxide solution — the sodium carbonate is a weaker alkali. Halogenoalkanes can be hydrolysed by strong alkali. Hence it is better to add sodium carbonate solution which neutralises the acid impurities but does not hydrolyse the product.

(e) Describe how the distillate is washed with aqueous sodium carbonate. [4]

> **(e)** Add the distillate to a separating funnel with a portion of aqueous sodium carbonate. ✓
>
> Stopper and shake, inverting the funnel and opening the tap to release the pressure. ✓
>
> Allow the separating funnel to stand until the layers settle and separate. Remove the stopper and open the tap, to run off the bottom layer into a beaker. Run off the second layer into a separate beaker and discard the aqueous layer. ✓
>
> Place the organic layer back into the separating funnel and repeat the process, using another portion of the aqueous solution. ✓

ⓔ Washing of an organic liquid to purify it is carried out in a separating funnel. Ensure that you mention safety precautions, such as release of pressure. This method must be learnt in detail.

(f) Without using densities, how would you determine which layer is the aqueous layer, during washing? [1]

ⓔ Remember that an aqueous solution is a solution in water.

> **(f)** Add water to the separating funnel. The layer that increases in size is the aqueous layer. ✓

Question 11

Ethanol is oxidised to ethanoic acid by heating under reflux using acidified potassium dichromate(VI) solution. What is the reason for heating under reflux? [1]

A to allow efficient mixing of the solutions

B to boil the mixture at a higher temperature

C to ensure even heating

D to prevent any substances escaping

The answer is D. ✓

ⓔ Reflux means the water-cooled condenser is in the vertical position. Mixing of the solutions could be achieved by shaking or agitation. The vertical condenser does not help with this. It does not allow the mixture to boil at a higher temperature. Impurities added would do this, and anti-bumping granules are needed to ensure even heating. The reason for reflux is to ensure that any volatile liquids do not escape. Hence answer D is correct.

Question 12

The melting point of paracetamol is 169°C. Different students A–D made paracetamol in the laboratory and recorded the melting point. Which student(s) made the purest paracetamol? Explain your answer.

[1]

Student	A	B	C	D
Melting point /°C	160–166	160–164	167–168	169–171

The answer is C as it has the narrowest range. ✓

ⓔ The purest solid will have the narrowest range, though if it is prepared in the lab it may still be lower than the true melting point.

Question 13

Methyl 3-nitrobenzoate is prepared by the following method:

Weigh out 2.7 g of methyl benzoate in a small conical flask and then dissolve in 5 cm^3 of concentrated sulfuric acid. When the solid has dissolved, cool the mixture in ice. Prepare the nitrating mixture by carefully adding 2 cm^3 of concentrated sulfuric acid to 2 cm^3 of concentrated nitric acid and then cool this mixture in ice. Add the nitrating mixture drop by drop to the solution of the methyl benzoate, stirring with a thermometer and keeping the temperature below 10°C. When the addition is complete, allow the mixture to stand at room temperature for 15 minutes. Pour the reaction mixture onto 25 g of crushed ice and stir until all the ice has melted and crystalline methyl 3-nitrobenzoate is formed. Filter the crystals using Büchner filtration, wash with cold water, recrystallise from ethanol and obtain the melting point.

Questions & Answers

(a) What homologous series does methyl benzoate belong to? What are the molecular and the empirical formulae of methyl benzoate? [3]

> **(a)** Ester ✓ $C_6H_5COOCH_3$ ✓ C_4H_4O ✓

ⓔ Methyl benzoate is an ester as it contains an ester bond –COOR. Remember that benzene is C_6H_6, but the H is substituted for $COOCH_3$, so the molecular formula is $C_6H_5COOCH_3$. To find the empirical formula, collect together each element $C_8H_8O_2$ and find the simplest ratio. In this case divide by 2.

(b) What conditions were used during this preparation to prevent further nitration of the product to the dinitro derivative? [1]

> **(b)** The temperature was kept low (less than 10°C) to prevent further nitration. ✓

(c) How do the conditions for the nitration of methyl benzoate differ from those for the nitration of benzene? [1]

> **(c)** Benzene is heated to 50°C. Methyl benzoate is cooled to less than 10°C. ✓

ⓔ You should know that benzene is nitrated by heating under reflux at 50°C.

(d) How do you expect the melting point of the impure sample to compare to the recrystallised sample? [2]

> **(d)** Lower ✓ wider range ✓

ⓔ Recrystallisation purifies a solid, so the impure sample should have a lower melting point and a wider range.

(e) Suggest ways in which the product may be lost during the experimental procedure. [2]

> **(e)** Some product is lost in purification steps/Some product is lost in transferring between vessels/Production of side products/Not all starting material reacted. Any 2 ✓✓

(f) Why were the crystals washed with cold water? [1]

> **(f)** To remove aqueous impurities, e.g. acid ✓

ⓔ Cold water will dissolve aqueous impurities, such as the concentrated acids.

Question 14

Isopropyl acetate is an ester with boiling point of 88°C. It can be prepared by heating acetic acid with isopropyl alcohol in the presence of a catalyst in a round-bottomed flask. The product is removed from the flask, purified and analysed.

OH
isopropyl alcohol

(a) (i) Write an equation for the equilibrium reaction between acetic acid (ethanoic acid) and isopropyl alcohol. [1]

ⓔ The product is an ester which has functional group –COO, made from ethanoic acid and the isopropyl alcohol. In this reaction water is removed.

(ii) Name the catalyst added to the flask. [1]

(iii) What is the name of the method of heating used, and what else should be added to the round-bottomed flask? [2]

(a) (i) $CH_3COOH + (CH_3)_2CHOH \rightleftharpoons CH_3COOCH(CH_3)_2 + H_2O$ ✓

(ii) Concentrated sulfuric acid ✓

(iii) Reflux ✓
Anti-bumping granules ✓

ⓔ In an esterification reaction the catalyst is usually a concentrated mineral acid, such as sulfuric acid. Organic liquids should be heated under reflux so that there is no loss of product and anti-bumping granules should be added to the flask to ensure even boiling.

(b) Give the IUPAC name for isopropyl alcohol. [1]

(b) propan-2-ol ✓

ⓔ There are three carbons in the longest chain, so prop is used, and the –OH group is on the second carbon.

(c) Assuming a 40% yield, what is the minimum mass of isopropyl alcohol required to produce 10.2 g of isopropyl acetate? [2]

ⓔ First calculate the moles of isopropyl acetate and then the mass of the alcohol. It is only a 40% yield, so to ensure 10.2 g of product, multiply the mass of alcohol by $\dfrac{100}{40}$.

(c) Moles of ester $= \dfrac{10.2}{102} = 0.1$
Moles of alcohol $= 0.1$
Mass of alcohol $= 0.1 \times 60 = 6\,g$
$6 \times \dfrac{100}{40} = 15\,g$ ✓✓

Questions & Answers

(d) Name the experimental technique used to remove the product from the flask. [1]

> **(d)** distillation ✓

@ The product will be mixed in the reaction flask with the catalyst and any unreacted reactants. All are liquids, so it is separated by distillation.

(e) Name a solution that can be added to remove acidic impurities from the crude product. [1]

> **(e)** sodium carbonate solution ✓

@ An aqueous solution of sodium carbonate will remove any acidic impurities, and it must be added in a separating funnel.

(f) Giving experimental details, describe how the crude product can be purified using anhydrous calcium chloride. [3]

> **(f)** Swirl with anhydrous calcium chloride ✓ until liquid goes clear. ✓
> Filter off the solid drying agent/Decant off the crude product. ✓

@ Anhydrous calcium chloride is a solid. It is a drying agent and can be added to the crude product in a flask, until the product changes from cloudy to clear, and it is then filtered off.

(g) Suggest how infrared spectroscopy could be used to show that the product did not contain any unreacted acetic acid or isopropyl alcohol. [1]

> **(g)** No –OH absorption ✓

@ Both acetic acid and the alcohol contain an –OH group. If they are not present, there should be no –OH absorption.

(h) State the integration pattern in the NMR spectrum of isopropyl acetate. [1]

@ There are three types of chemically equivalent hydrogen atoms. The integration gives the ratio of each type of chemically equivalent hydrogen.

> **(h)** The integration pattern is $3:1:6$. ✓

Question 15

Methanal is gaseous at room temperature, but can be obtained as a solution which contains 37.0% methanal by mass. Methanal causes severe irritation of the nasal system. It can be reacted with excess aqueous 2,4-dinitrophenylhydrazine to form a hydrazone.

(a) Write the equation for the reaction of methanal with 2,4-dinitrophenylhydrazine. [2]

(a)

✓✓

(b) Calculate the mass of methanal solution needed to form 1.4 g of the
2,4-dinitrophenylhydrazone assuming a 95% yield. [4]

ℯ First find the relative molecular masses: $M_r\ CH_2O = 12 + 2 + 16 = 30$ and
$M_r\ C_7H_6N_4O_4 = 210$. Then find the moles of hydrazone, moles of methanal by
ratio and the mass of methanal. The methanal solution contains 37% methanal
by mass.

(b) Moles $= \dfrac{1.4}{210} = 0.0066667$ ✓

Mass of methanal $= 0.0066667 \times 30 = 0.2000\,g$ ✓

The methanal solution contains 37.0% methanal by mass, so the mass
needed is $\dfrac{100}{37} \times 0.2 = 0.541\,g$ ✓

For a 95% yield $0.541 \times \dfrac{100}{95} = 0.57\,g$ ✓

(c) Explain the safety precaution that needs to be followed during the preparation. [1]

(c) Use a fume cupboard as methanal irritates the nose. ✓

(d) The 2,4-dinitrophenylhydrazone is formed as an orange precipitate, which is
collected by suction filtration using a Büchner flask.

(i) Explain how 'Büchner filtration' is carried out. [3]
(ii) State why it is used in preference to normal filtration. [2]

(d) (i) Place filter paper into Büchner funnel. ✓
Funnel placed into Büchner flask. ✓
Suck the air through/attach to pump. ✓

(ii) It is faster. ✓
It produces drier crystals. ✓

(e) The solvent used to recrystallise the 2,4dinitrophenylhydrazone is ethanol.
Explain why the 2,4-dinitrophenylhydrazone is soluble in ethanol and not
in hexane. [2]

(e) Hydrazone can form hydrogen bonds with ethanol. ✓

Hexane is non-polar and cannot form hydrogen bonds with hydrazine. ✓

(f) Describe, giving experimental details, how the 2,4-dinitrophenylhydrazone is recrystallised. [4]

🅔 Note that the stem of the question states that the solvent is ethanol, so make sure you mention ethanol in your answer.

> (f) Dissolve in the minimum volume of hot ethanol. ✓
>
> Filter when hot. ✓
>
> Allow to cool and crystallise. ✓
>
> Filter under reduced pressure. ✓

Question 16

Some tests were carried out an organic liquid B, which has the *empirical* formula C_2H_4O. The results are shown in the table below.

Test	Description	Observation
1	Add $2\,cm^3$ of deionised water to $2\,cm^3$ of B in a test tube.	One layer forms
2	Add 10 drops of B to $2\,cm^3$ of acidified potassium dichromate solution in a test tube. Place the test tube in a hot water bath.	Solution remains orange
3	Place $5\,cm^3$ of B in a boiling tube. Add $5\,cm^3$ of ethanol, and then $1\,cm^3$ of concentrated sulfuric acid. Heat the boiling tube in a water bath. Cautiously smell the contents of the boiling tube.	Sweet smell
4	Add a spatula measure of sodium carbonate to $2\,cm^3$ of B in a test tube.	Effervescence

(a) What can be deduced from test 1? [1]

> (a) B is miscible with water/–OH group present/B can hydrogen bond with water. ✓

🅔 If an organic substance forms one layer with water, it means it is miscible. This is due to hydrogen bonds forming between the organic compound and the water, due to the organic compound possibly containing an –OH group.

(b) What can be deduced from test 2? [1]

> (b) B is not a primary or secondary alcohol or aldehyde. ✓

🅔 Acidified potassium dichromate changes colour from orange to green when warmed with an aldehyde, primary or secondary alcohol. There is no colour change and so B is not one of these, but may be a tertiary alcohol or a carboxylic acid.

(c) What can be deduced from test 3? [1]

> (c) B may be a carboxylic acid/An ester has formed. ✓

ⓔ A sweet-smelling substance is the ester formed when the ethanol reacts with B in the presence of a catalyst of concentrated sulfuric acid. It is likely that B is a carboxylic acid.

(d) What can be deduced from test 4? [1]

(d) B is a carboxylic acid. ✓

(e) Describe a test that could be used to test for the gas produced in test 4. [2]

(e) Bubble the gas into limewater. ✓ It changes from colourless to cloudy. ✓

(f) The mass spectrum of B is shown below.

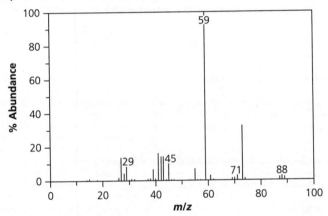

Draw a displayed structure for B. [1]

ⓔ B is a carboxylic acid. It has empirical formula C_2H_4O. The mass spectrum shows that it has a relative molecular mass of 88, so its molecular formula is $C_4H_8O_2$.

(f)

H—C—C—C—C—O—H   *or*   H—C—C—C—O—H

✓

(g) Identify the species responsible for the base peak in the spectrum above. [1]

(g) $CH_2COOH^+/CH_2COOH^+/C_2H_3O_2^+$ ✓

Question 17

Describe chemical tests that you could carry out in test tubes to distinguish between compounds X, Y and Z.

Include appropriate reagents and any relevant observations. Also include equations showing structures for the organic compounds involved. [6]

X   Y   Z

Warm all solutions with Tollens' reagent and a silver mirror is formed with Y or warm with Fehling's solution and a red ppt is formed with Y. ✓

Then warm fresh samples of X and Z with acidified potassium dichromate(vi). ✓

Orange solution changes to green in X which is a primary alcohol. ✓

✓

Orange solution stays orange in Z (tertiary alcohol). ✓

ⓔ Compound Y has a carbonyl group at the end of the chain and is an aldehyde, which will react with Tollens' reagent or Fehling's solution. X and Z both have alcohol groups but X is a primary alcohol so it will be oxidised by acidified potassium dichromate(vi), and Z is a tertiary alcohol and cannot be oxidised. The order of reactions can be different.

Question 18

In an experiment to investigate cells, a student set up a cell between a silver half-cell and a zinc half-cell under standard conditions. The standard redox potentials for each half-cell are given below:

$$Ag^+(aq) + e^- \rightarrow Ag(s) \qquad E^\ominus = +0.80\,V$$
$$Zn^{2+}(aq) + 2e^- \rightarrow Zn(s) \qquad E^\ominus = -0.76\,V$$

(a) Outline the experimental setup that could be used in the laboratory to measure the standard cell potential of the cell. In your answer you should include a diagram of the setup and the standard conditions required. [4]

(a)

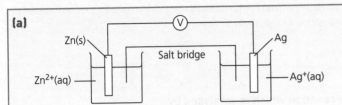

One mark for each of the following labels:

Ag(s)$^+$ 1 mol dm^{-3} Ag$^+$(aq) ✓

Zn(s)$^+$ 1 mol dm^{-3} Zn^{2+}(aq) ✓

298 K/25°C and 100 kPa/101 kPa ✓

Salt bridge voltmeter and wires to complete circuit ✓

ℯ To experimentally determine a standard cell potential, each half-cell must consist of the metal dipping into a 1 mol dm^{-3} solution of the metal ions. A high-resistance voltmeter is used and a salt bridge.

(b) Calculate the standard cell potential of the cell. [1]

(b) $E^{\ominus}_{cell}$ = E(reduction reaction) – E(oxidation reaction)

$E^{\ominus}_{cell}$ = +0.80 – (–0.76) = +1.56 V ✓

ℯ The zinc loses electrons and forms zinc ions in an oxidation reaction.

(c) Indicate the direction of flow of electrons in the external circuit. [1]

(c) The electrons move from the zinc half-cell through the external circuit to the silver half-cell. ✓

(d) Why is a salt bridge made of filter paper saturated with potassium chloride solution not effective as a salt bridge in this reaction? [1]

ℯ Think about the ions that are present in the aqueous solutions in the beaker — silver ions and zinc ions.

(d) The chloride ions would react with the silver ions, precipitating silver chloride. ✓

(e) Suggest a change to one half-cell that would increase the overall cell potential. [1]

(e) Decreasing the concentration of zinc ions/ increasing the concentration of silver ions ✓

@ By decreasing the concentration of the zinc ions, the position of the zinc equilibrium will shift to the left, making the electrode potential more negative. This will increase the cell potential. Equally, increasing the concentration of the silver ions will shift the position of the silver equilibrium to the right, making the electrode potential more positive, also increasing the magnitude of the cell potential.

Question 19

Bromine reacts with aqueous methanoic acid in a reaction which is catalysed by hydrogen ions:

$$Br_2(aq) + HCOOH(aq) \xrightarrow{H^+(aq)} 2Br^-(aq) + 2H^+(aq) + CO_2(g)$$

The rate of the reaction can be followed using a colorimeter. The methanoic acid is present in a large excess.

(a) Why can colorimetry be used to investigate the rate of this reaction?   [1]

> **(a)** The orange colour of bromine fades ✓ as the reaction proceeds and colourless bromide ions are produced.

@ Colorimetry is used to investigate the rate of a reaction if a reactant or product is coloured. Look carefully at the reaction equation and determine if there is a coloured substance present — an aqueous solution of bromine is orange. As the reaction proceeds, the colour of the bromine fades and can be monitored by colorimetry.

(b) Describe how colorimetry could be used to determine the order of reaction with respect to bromine.   [4]

> **(b)** Measure absorbance against time. ✓
>
> Use calibration curve of known concentrations of bromine to convert absorbance to concentration. ✓
>
> Plot bromine concentration against time and determine order from shape/ Find half-lives or find gradient of tangents at points and plot rate against bromine concentration and determine order from shape of graph. ✓

@ From experimental data there are different ways of determining the order — plotting $[Br_2]$ against time and using the shape of the graph to determine the order with respect to Br_2 (and verifying first order using half-life calculations), or taking tangents to the $[Br_2]$ against time graph at various concentrations and plotting a rate versus $[Br_2]$ graph to identify the order.

(c) Suggest an alternative method for investigating the rate of this reaction.   [1]

@ Again look carefully at the reaction equation. Carbon dioxide gas is also produced.

> **(c)** Use a gas syringe to measure the volume of gas produced against time. ✓

(d) Suggest a suitable chemical to use as the catalyst for the reaction. [1]

> **(d)** Any strong acid such as hydrochloric ✓

(e) Explain the purpose of adding a large excess of methanoic acid. [2]

> **(e)** The concentration of methanoic acid is constant. ✓
>
> The effect of changing the concentration of bromine on the rate can be determined. ✓

ℯ Adding a large excess of methanoic acid means that its concentration stays constant during the reaction so the effect of changing the bromine concentration on the rate can be measured.

(f) Use the results below to plot a graph of bromine concentration against time. [3]

Time/s	$[Br_2]/10^{-3} \, mol \, dm^{-3}$	Time/s	$[Br_2]/10^{-3} \, mol \, dm^{-3}$
0	10.0	180	5.3
10	9.0	240	4.4
30	8.1	360	2.8
90	6.7	480	2.0
120	6.2	600	1.3

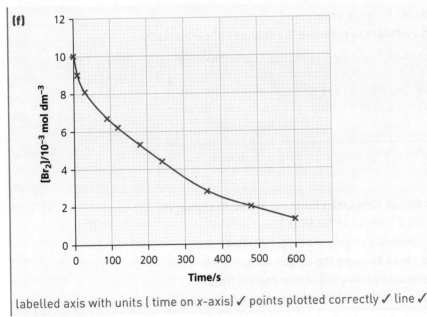

(f)

labelled axis with units (time on x-axis) ✓ points plotted correctly ✓ line ✓

(g) Determine two values for half-lives for the reaction from your concentration–time graph and determine the order of the reaction. [2]

> **(g)** All half-lives are around 200 ✓ first order ✓

ⓔ From the concentration–time graph the shape is similar to that of a first-order reaction. To find the half-life, determine the time taken for the concentration to fall from 1.3 to 0.65 etc.

Question 20

(a) The apparatus shown in the figure below was used to measure the rate of a reaction for a precipitation reaction of two solutions. $20\,cm^3$ of solution A was measured and added to a conical flask, and $20\,cm^3$ of solution B added and a timer started. The timer was stopped when it was no longer possible to see the cross on the paper. The experiment was repeated.

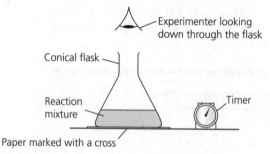

What is likely to decrease the accuracy of the experiment?   [1]

A   Rinsing the flask with the solution A before each new experiment.

B   Stirring the solution throughout each experiment.

C   Using the same cross on paper for each experiment.

D   Using different measuring cylinders to measure the volumes of each solution.

ⓔ In this practical there should be some controlled variables. The same cross should be used, so it is of the same darkness in all experiments, and if the reactants are stirred they should be stirred in all experiments. Using different measuring cylinders increases accuracy as it prevents precipitation occurring in the cylinder, if some drops of the previous solution are left in the cylinder.

> **(a)** The answer is A — rinsing the flask with solution A before each new experiment. ✓

(b) Which statement is correct about the time taken for the cross to disappear if the experiment is repeated using a larger conical flask?   [1]

A   The time taken will not be affected by using the larger conical flask.

B   The time taken will be decreased by using the larger conical flask.

C   The time taken will be increased by using the larger conical flask.

D   It is impossible to predict how the time taken will be affected by using the larger conical flask.

> **(b)** The answer is C — the time taken will be increased by using the larger conical flask. ✓

ⓔ Using a larger flask means that the precipitation will not form as thick a covering and so it will take longer for the cross to be obscured.

Question 21

The equations for reaction A and B are:

Reaction A: $2H_2O_2(aq) \rightarrow 2H_2O(l) + O_2(g)$

Reaction B: $CH_3COCH_3(aq) + I_2(aq) \rightarrow CH_3COCH_2I(aq) + H^+(aq) + I^-(aq)$

Choose which of the following apparatus could be used to continuously monitor the rate of (a) reaction A, (b) reaction B.

[2]

balance   colorimeter   gas syringe   pH meter

> (a) Reaction A: balance and gas syringe ✓
>
> (b) Reaction B: colorimeter and pH meter ✓

e For reaction A a gas is produced, so a gas syringe or a balance could be used. Hydrogen peroxide and water are both colourless so a colorimeter could not be used.

For reaction B iodine is coloured, so a colorimeter could be used. In addition, hydrogen ions are generated so a pH meter could be used.

Question 22

An inorganic compound contains a cation which is either NH_4^+, Fe^{3+} or Cr^{3+} and an anion which is Cl^-, SO_4^{2-} or CO_3^{2-}.

Plan a series of tests that you could carry out on the samples to identify the ionic compound. Your tests should produce at least one positive result for each ion.

For each test:
- include details of reagents used, relevant observations and equations
- explain how your observations allow the ions to be identified

You may include flowcharts or tables in your answer.

[8]

e The anion tests should be carried out in the order below. Otherwise dilute nitric acid can be added before the silver nitrate, and barium nitrate. Make sure you read the question carefully. You need to include reagents, observations and equations. It is best to give ionic equations for ion tests.

For the cation tests:

Add aqueous sodium hydroxide. ✓

  Fe^{3+} brown ppt: $Fe^{3+} + 2OH^- \rightarrow Fe(OH)_3$ ✓
  Cu^{2+} blue ppt: $Cu^{2+} + 2OH^- \rightarrow Cu(OH)_2$ ✓

Warm a sample with sodium hydroxide solution and test the gas released with moist red litmus and it should turn blue, indicating that alkaline ammonia has been produced.
  $NH_4^+ + OH^- \rightarrow NH_3 + H_2O$

For the anion tests:

This should be carried out in order. ✓

Add nitric acid. If there is effervescence, it is a carbonate:
$$CO_3^{2-} + 2H^+ \rightarrow H_2O + CO_2 ✓$$

Add barium nitrate. If there is a white ppt, then it is a sulfate:
$$Ba^{2+} + SO_4^{2-} \rightarrow BaSO_4 ✓$$

Add silver nitrate. If there is a white ppt, it is a chloride. Add some dilute aqueous ammonia and the ppt should dissolve: $Ag^+ + Cl^- \rightarrow AgCl$ ✓

Question 23

A pH meter was used to investigate the effect of dilution on the dissociation of ethanoic acid. The solutions shown in the table were prepared by diluting a $0.10 \, mol \, dm^{-3}$ solution of the acid. A pH meter was calibrated and the pH of each solution measured, starting with the least concentrated.

Concentration of ethanoic acid/$mol \, dm^{-3}$	pH reading	Calculated pH for solutions of hydrochloric acid of the same concentration
0.00010	4.2	
0.0010	3.5	
0.010	3.0	
0.10	2.7	1.0

(a) Describe how a $0.010 \, mol \, dm^{-3}$ solution of ethanoic acid was prepared from the $0.10 \, mol \, dm^{-3}$ solution. [3]

ⓔ Dilutions are made up in a volumetric flask. The dilution factor is ×10, so $25.0 \, cm^3$ is diluted to $250.0 \, cm^3$.

> (a) Rinse a pipette with the $0.010 \, mol \, dm^{-3}$ ethanoic acid and then use a safety pipette filler with the pipette to transfer $25.0 \, cm^3$ of the solution ✓ into a $250 \, cm^3$ volumetric flask ✓. Make up with deionised water until the bottom of the meniscus is on the zero mark at eye level ✓.

(b) The water used for the dilutions had been boiled and then allowed to cool to room temperature. Explain why this improved the accuracy of the measurements. [2]

ⓔ Deionised water contains acidic carbon dioxide. This may affect the pH of water.

> (b) Boiling removes dissolved carbon dioxide ✓ which is acidic and may affect the pH of the water ✓.

(c) Explain why it was necessary to calibrate the pH meter. [1]

(c) After storage, a pH meter does not give accurate readings because the glass electrode in the pH meter does not give a reproducible emf over longer periods of time. ✓

(d) Why did the students measure the pH of the most dilute solution first and then work up to the more concentrated solutions? [1]

(d) It is difficult to ensure that all of the hydrogen ions had been rinsed away with deionised water and this could affect the result ✓.

e The pH meter responds to the presence of hydrogen ions. If the most concentrated solution were measured first, it would be difficult to ensure that all the hydrogen ions had been rinsed away with deionised water and this could affect the result.

(e) Describe how the pH was measured. [2]

(e) Place the pH probe in the solution, record the value to one decimal place. ✓

Remove the probe and wash with deionised water. ✓

(f) Calculate the pH values for hydrochloric acid which are missing from the table. [1]

e Hydrochloric acid is a strong acid so use the equation pH = $-\log_{10}[H^+]$

(f) 4.0, 2.0, 1.0 ✓

(g) What do the differences in pH of each acid at each concentration show about the effect of dilution on the degree of dissociation of ethanoic acid? [2]

(g) The values get closer as the acids get more dilute. ✓

The degree of dissociation increases with dilution. ✓

e The pH values of ethanoic acid get closer to the pH values of hydrochloric acid as the acids get more dilute, showing that the degree of dissociation of the weak acid increases as it is diluted.

Question 24

Which one of the following techniques could be used to accurately determine the percentage iron(II) ion content of an iron sulfate tablet? [1]

A thin layer chromatography

B melting point determination

C addition of a neutral solution of iron(III) chloride

D titration with potassium manganate(VII).

Questions & Answers

> The answer is D. ✓

ⓔ Thin layer chromatography does not determine quantitatively the iron(II) content. Melting point determination shows purity, not the percentage content.

Question 25

A student carried out an experiment to determine which of a chloroalkane, bromoalkane or iodoalkane hydrolysed fastest when heated with silver nitrate in ethanol.

(a) Describe an experimental method that he could use. Ensure it is a fair test. [6]

> **(a)** Any two fair test marks:
>
> Use haloalkanes with the same chain length. ✓
>
> Equal amounts of ethanol, silver nitrate mixture in each test tube and equal volumes of haloalkane. ✓
>
> All tubes at same temperature. ✓
>
> Any four method marks:
>
> **1** Place (equal volumes) of each haloalkane in separate test tubes and place in a water bath at 50°C. ✓
>
> **2** Heat a test tube of a mixture of ethanol and aqueous silver nitrate in the same water bath. ✓
>
> **3** When all tubes have reached the same temperature, add (equal volumes) of the mixture to each haloalkane and start the clock. ✓
>
> **4** Time how long it takes for each precipitate to form. ✓

ⓔ In this reaction the haloalkane hydrolyses and halide ions are produced. Hence, as the haloalkane is hydrolysed and the reaction proceeds, halide ions are produced and a precipitate of silver halide gradually appears when the halide ion reacts with the silver nitrate. The rate of precipitation is a measure of the rate of hydrolysis. You may name the haloalkanes used, for example chlorobutane, bromobutane and iodobutane, or refer to the fact that they have the same chain length, for a fair test. Note that a water bath must be used to heat due to the flammable nature of ethanol.

(b) State and explain the results, in terms of ease of hydrolysis of the haloalkanes, which you would expect in this experiment. [2]

> **(b)** The iodoalkane forms a precipitate first, then the bromoalkane, then the chloroalkane. ✓
>
> The C–I bond energy is smaller than the C–Br and C–Cl bond energies. ✓

(c) Describe how you would identify the chloroalkane, bromoalkane and iodoalkane in this experiment. [1]

(c) White ppt chloroalkane, cream ppt bromoalkane, yellow ppt iodoalkane ✓

(d) The densities and the boiling points of haloethanes are listed in the table below.

Haloethane	Density/g cm^{-3}	Boiling point/°C
Chloroethane	0.898	12
Bromoethane	1.461	38
Iodoethane	1.936	72

(i) Suggest why there is an increase in density from bromoethane to iodoethane. [1]

(d) (i) Increase in mass of halogen atom ✓

ℯ Comparing the haloethanes, the difference between them is due to the different atoms — they have different masses. The increase in density is because the halogen atom increases in mass.

(ii) Suggest why there is an increase in boiling point from chloroethane to iodoethane. [2]

(ii) More electrons/greater M_r in iodoethane ✓
Stronger van der Waals forces between molecules ✓

ℯ Boiling point can be explained by looking at the intermolecular forces. Due to the larger relative molecular mass in iodoethane, there are stronger van der Waals forces and more energy is needed to break these.

(iii) If all three haloethanes were present in a container at room temperature, suggest how you would separate and obtain each haloethane. [2]

(iii) Chlorethane is a gas and must be removed/Cool to liquefy. ✓
Fractional distillation collect at boiling points. ✓

ℯ Look carefully at the data to help you interpret this. At room temperature the chloroethane is a gas and the other two are liquids.

Knowledge check answers

1 Down the sink is safe as the product is a neutral solution of sodium chloride.

2 $7.9\,g\,cm^{-3}$

3 To ensures all the magnesium reacts

4 Do not touch hot evaporating basin/allow to cool before weighing/use tongs to lift evaporating basin/wear gloves when lifting evaporating basin

5 To ensure all the water of crystallisation is removed.

6 overall uncertainty = $2 \times 0.005\,g$
mass lost = $1.24\,g$
percentage uncertainty in mass loss
$= (2 \times 0.005 \times 100)/1.24 = 0.81\%$

7 percentage error = $\frac{0.01}{0.12} \times 100 = 8.3\%$

percentage error = $\frac{1}{120} \times 100 = 0.8\%$

8 It absorbs water from the atmosphere.

9 Pink to colourless

10 $CH_4 + 2O_2 \rightarrow CO_2 + 2H_2O$

11 Mass of spirit burner + methanol, mass of spirit burner + methanol at end, mass of water

12 To prevent draughts and so prevent heat loss

13 Carbon monoxide, carbon (soot)

14 Only hydrogen and hydroxide ions are involved in the reaction $H^+ + OH^- \rightarrow H_2O$, so the nature of the acid and alkali does not matter.

15 Warm the sample with sodium hydroxide solution. A pungent-smelling gas is released which changes moist indicator paper to blue and gives white fumes with a glass rod dipped in concentrated hydrochloric acid.

16 White precipitate

17 $Cu^{2+} + 2OH^- \rightarrow Cu(OH)_2$

18 A separating funnel, showing two liquid layers. Using the density, the oil is the top layer.

19 Anhydrous sodium sulfate/anhydrous magnesium sulfate/anhydrous calcium chloride

20 Ethanoic acid is soluble in water.

21 Reflux

22 Recrystallisation

23 Lower than that of the pure substance and with a wider range

24 % yield = $\frac{actual\ yield}{theoretical\ yield} \times 100$

25 The bromine adds on to the carbon–carbon double bond.

26 Silver ion $Ag^+ + e^- \rightarrow Ag$

27 $2CH_3CH_2COOH + Na_2CO_3 \rightarrow 2CH_3CH_2COONa + H_2O + CO_2$
$2CH_3CH_2COOH + Mg \rightarrow (CH_3CH_2COO)_2Mg + H_2O + CO_2$
effervescence

28 temperature of 298 K, gases at pressure of 100 kPa, solutions at a concentration of $1.0\,mol\,dm^{-3}$

29 a $Mg + CuSO_4 \rightarrow MgSO_4 + Cu$
  $Mg \rightarrow Mg^{2+} +2e$ Mg is oxidised
  $Cu^{2+} + 2e^- \rightarrow Cu$ Cu^{2+} is reduced

 b $Cl_2 + 2KBr \rightarrow Br_2 + 2KCl$
  $Cl_2 \rightarrow 2Cl^- + 2e^-$ Cl_2 is oxidised
  $2Br^- \rightarrow Br_2 + 2e^-$ Br^- is reduced

30 $50.0\,cm^3$ burette

31 colorimeter

32 The time taken for the concentration of one of the reactants to fall by half.

33 first order, second order, $mol^{-2}dm^6s^{-1}$

34 s^{-1}

35 rate = $k[H_2] [NO]^2$

36 A strong acid is one that dissociates into ions completely, in solution. A weak acid is one that partially dissociates in solution. A buffer is a substance that resists changes in pH when small amounts of acid or alkali are added or when diluted.

Note: **bold** page numbers indicate defined terms.

Index